Are Women Achieving Equity in Chemistry?

ACS SYMPOSIUM SERIES **929**

Are Women Achieving Equity in Chemistry?

Dissolving Disparity and Catalyzing Change

Cecilia H. Marzabadi, Editor
Seton Hall University

Valerie J. Kuck, Editor
Seton Hall University

Susan A. Nolan, Editor
Seton Hall University

Janine P. Buckner, Editor
Seton Hall University

**Sponsored by the
ACS Women Chemists' Committee, the Society Committee on Education, and the ACS Divisions of Chemical Education, Inc., and Professional Relations**

American Chemical Society, Washington, DC

Library of Congress Cataloging-in-Publication Data

American Chemical Society. Meeting (226th : 2003 : New York, N.Y.)

Are women achieving equity in chemistry? dissolving disparity and catalyzing change / Cecilia H. Marzabadi ... [et al.] ; ; sponsored by the ACS Women Chemists Committee ... [et al.].

p. cm.—(ACS symposium series ; 929)

Includes bibliographical references and indexes.

ISBN 13: 978–0–8412–3950–0 (alk. paper)

1. Women in chemistry—Research—United States—Congresses. 2. Women chemists—Research—United States—Congresses.

I. Marzabadi, Cecilia H. II. American Chemical Society. Women Chemists Committee. III. Title. IV. Series.

QD20.A44 2003
540.820973—dc22 2005053646

The paper used in this publication meets the minimum requirements of American National Standard for Information Sciences—Permanence of Paper for Printed Library Materials, ANSI Z39.48–1984.

Distributed by Oxford University Press

ISBN 10: 0–8412–3950–9

PRINTED IN THE UNITED STATES OF AMERICA

10 9 8 7 6 5 4 3 2

Foreword

The ACS Symposium Series was first published in 1974 to provide a mechanism for publishing symposia quickly in book form. The purpose of the series is to publish timely, comprehensive books developed from ACS sponsored symposia based on current scientific research. Occasionally, books are developed from symposia sponsored by other organizations when the topic is of keen interest to the chemistry audience.

Before agreeing to publish a book, the proposed table of contents is reviewed for appropriate and comprehensive coverage and for interest to the audience. Some papers may be excluded to better focus the book; others may be added to provide comprehensiveness. When appropriate, overview or introductory chapters are added. Drafts of chapters are peer-reviewed prior to final acceptance or rejection, and manuscripts are prepared in camera-ready format.

As a rule, only original research papers and original review papers are included in the volumes. Verbatim reproductions of previously published papers are not accepted.

ACS Books Department

Contents

Indexes

Preface

This volume is a result of a day-long symposium held at the 226th National American Chemical Society (ACS) Meeting in New York City in the fall of 2003. The symposium focused on women in academe and the barriers that they are currently facing not only in the United States but also in Europe. To increase the breadth of the symposium, social scientists as well as physical scientists were included as speakers.

The purpose of the symposium was to discuss the reasons for the low representation of women on chemistry faculty at research institutions and to propose ways to increase women's participation and retention in academe. In light of declining enrollments in chemistry departments in the United States, both by males and by non-U.S. citizens, it is important that efforts are made to begin to incorporate the diversity of the available U.S. workforce into science, technology, engineering, and mathematics (STEM) fields, if we are to remain globally competitive in these disciplines. Women are a particularly attractive target for recruitment because they currently receive more than 50% of the bachelor's degrees and more than 33% of the doctoral degrees each year in these disciplines. Yet, the percentage of female chemistry faculty available to mentor students has failed to rise proportionately during the past several decades. Currently, most research departments have fewer than 15% of their faculty comprised by women.

Tangible evidence for the reasons women opt out of these careers has been lacking. This book brings together a collection of archival and survey data that documents the reasons for this disparity. The chapters in this volume look at the different phases in a woman's academic career path where exit from the "pipeline" are likely to occur. For example, the graduate and postdoctoral training environments of the future scientist are examined and the outcomes of these experiences on women's subsequent participation in chemistry are assessed. Societal practices that impede or discourage women from pursuing jobs in academe within STEM fields also are described.

Furthermore, this volume discusses the factors that affect a woman's advancement and retention in an academic job. The perceptions of female faculty recipients of prestigious awards and professorships are presented in order to understand the difficulties that women are experiencing. In addition, the reasons why women fail to thrive in academe are explored as well as why women leave science fields altogether. The similarities in inequities that academic women are facing in the United States and Germany also are discussed. Successful programs and practices to encourage the promotion and retention of women in chemistry are presented. The major contribution of this effort is to clearly show that the struggle for equality in chemistry is ongoing and is international.

This symposium was cosponsored by the ACS Women Chemists' Committee, Society Committee on Education, and the ACS Divisions of Professional Relations and Chemical Education, Inc. It was generously funded by the National Science Foundation (grant CHE–0341126).

Cecilia H. Marzabadi
Department of Chemistry and Biochemistry
Elizabeth Ann Seton Center for Women's Studies
Seton Hall University
South Orange, NJ 07079

Are Women Achieving Equity in Chemistry?

Chapter 1

The Status of Women Chemists in Academe: An Introduction

Janine P. Buckner

Department of Psychology and the Center for Women's Studies, Seton Hall University, 400 South Orange Avenue, South Orange, NJ 07079 (email: buckneja@shu.edu)

An overview of the status of women in academic chemistry is presented. Perspectives by both physical and social scientists on the reasons for women's under-employent and low retention in tenure track positions at US research universities are summarized. In comparison, the situation for women in Europe is also described. Within the context of these discussions, best practices to increase diversity in the physical sciences can be derived.

The rate of doctoral degree attainment for women in the field of chemistry has been shown to be moving in a consistently positive direction over the last several decades. Yet, despite the steady rise in the pool of available female candidates for academic employment, the number of women actually obtaining tenure-track faculty positions in chemistry departments at doctoral institutions has not increased to the same degree. This is particularly true at the elite universities. As the title of this book suggests, the motivation for the inquiries in this volume is to question whether and how women can in fact achieve the equity in chemistry they are seeking, and to document the equity they have achieved at some levels. This issue is not limited to women once they have secured employment in academic professorate roles, but extends to pre-employment training and experiences as well; that is, are women receiving the same kinds of education, training, mentoring, support, and opportunities for employment and advancement that their male peers are receiving? Indeed, recent work by my colleagues and I suggests that women sometimes perceive very disparate experiences within the field of chemistry than do their male peers and colleagues (*1, 2, 3, 4*). Given these data, and the training and employment climates precipitated by this gendered treatment, we ask what practices should be instituted to dissolve the gap between men's and women's experiences, in order to stimulate lasting, effective changes.

The papers presented in this book emanate from a recent symposium in which social and physical scientists expounded upon these issues and proffered possible remedies to correct this situation. The authors in this book represent a broad variety of perspectives, and draw from the experiences of women in chemistry and other physical sciences; from data obtained by academic researchers in social science disciplines such as psychology and sociology; from women in leadership positions at federal as well as private funding foundations; and from women across the globe, not only from within the United States. Taken in total, these perspectives weave together a tapestry of experiences that suggest that women are not perceived or treated by others as being equivalent to men in terms of the support and encouragement that they receive, that the amount and quality of education and training experiences offered to them is not equivalent to that received by men, and that the funding, hiring, and formal and informal advocacy provided on their behalf is less than that given to their male colleagues.

Given that this book represents scientific conclusions drawn from data collated by scientific methodologies, we would be remiss if we merely described the problematic situation and explained why such differences between women and men occur, but did not fulfill our obligation to then predict future patterns and attempt to modify possible negative outcomes for future scientists. Thus, throughout the contributions that comprise this volume, each of the authors highlights possible practices that may alleviate the current conditions and improve the prospects for male and female students who strive to join the ranks of academe. Individuals considering careers in the physical sciences—and those who mentor these individuals—would be well served to consider what the data do and do not say on the issue of disparity between men and women in their fields.

On the Status of Women Academicians

Although chemists and other scientists have become more aware of the disparity that exists between women and men in hiring and promotion in academe, a comprehensive understanding of the reasons and practices contributing to these gender inequities is not commonly shared among colleagues. One reason for this lack of awareness is that dialogue about these issues and dissemination of ideas stimulated by research are not broad enough to reach the general population of scientists. In recent years, however, social scientists have begun to investigate such disparities. In particular, researchers have examined how the institutional environment and practices have negatively impacted the participation of women in the physical sciences and engineering (collectively known as the Science, Technology, Engineering, and Mathematical fields, or STEM). This recent work suggests that women are greatly underrepresented in the STEM workforce (*5, 6, 7, 8, 9, 10*). In our own recent national survey of doctoral recipients (1988-1992) from top ranked institutions in chemistry (*1, 3, 4, 11*) we discovered differential patterns in female and male participants' perceptions of their graduate education, employment choices, and career continuity; moreover, we found differences in women's and men's projections of and expectations for academic employment. A careful examination of the criteria named as motivators for these career choices by this "elite pool" of respondents also reveals gender patterns. These and other data (e.g., *5, 12, 13*), placed particular emphasis upon some of the major factors that appear to be related to academic development and the disparate early career paths of women and men, such as the role of mentoring (*6*)

To our knowledge, this book is one of the first to examine perceptions of the graduates themselves with respect to their training and career trajectories. In the following pages, we will present, for the first time, a quantitative analysis of new aspects of our own dataset and discussions of findings from other fields. This discussion will consider the numbers of job interviews, offers, and acceptances received by the elite group of male and female chemistry doctoral recipients. We also will explore reasons and criteria cited by this select group of individuals for their choice of graduate school, dissertation advisor, post-doctoral position, and career path. We especially find important several statistical differences in student and post-doctoral perceptions of mentoring relationships. These additional data lend more support to the hypothesis that individuals' own perceptions of lifestyle and personal abilities, combined with mentoring experiences, access to resources, and networking relationships, can have a very strong effect on the expected career trajectories of up-and-coming academicians.

Although we find these data to be significant and important, data collected by other scientists corroborate these patterns of disparity in academia, and are equally compelling. Combined, the studies and discussions comprising this book present a rather strong case for continued gender inequities in many areas of the academic world. In STEM fields more generally—not just in chemistry—the path from graduate school through post-doctoral positions to academic careers continues to lose women every step of the way.

In the following chapters we present statistics from a variety of disciplines that demonstrate differential success rates in graduate school, post- doctoral attainment, and subsequent academic employment for men and women in STEM fields. Specific to the field of chemistry, it appears that women are held to different standards of evaluation and qualification than are men; for instance, women are required to complete more prestigious post-doctoral positions than are men in order to obtain top-tier academic positions. Though not as pronounced, gender dissimilarities also are evident in faculty compositions of top-ranked institutions in other STEM disciplines. As we have reported elsewhere, more men are hired by the top schools than are women. In effect, it appears that in considering doctorates from an elite school, men seem to be preferentially hired (*3*).

The motivation for this book is to present a detailed analysis of the factors influencing the under-employment and low retention of women in academe. The perspectives of the disparity between women and men in chemistry emanate from several disciplines—from the physical sciences to the social sciences. Yet, the diversity represented by the authors of this text converges upon a singular goal: to elucidate why disparate patterns exist in the education, training, mentoring, and employment opportunities afforded to female and male colleagues of both junior and senior status. In order to achieve departments that have faculties with the same gender distribution as the student bodies that they educate, it is critical that an understanding of the training and career issues that women in the physical sciences and engineering face be obtained.

The impetus for this book, therefore, is the need to present these issues. The contents wherein emerged out of a one-day symposium entitled, "Dissolving Disparity, Catalyzing Change: Are Women Achieving Equity in Chemistry?" which took place in New York City during the 226th national meeting of the American Chemical Society (ACS), September 7-11, 2003. The symposium was sponsored by several committees of the American Chemical Society, including the Committees on Women Chemists (WCC), on Professional Chemists (PROF), and Education (SOCED), and the Division of Chemical Education (DIVCHED). The symposium was funded by the National Science Foundation (CHE-0341126).

Though the symposium was not limited strictly to research in chemistry, our goal was to delineate the difficulties that women in the physical sciences and engineering are encountering in academe. This is of utmost importance because, with this knowledge, chemists can have a far better understanding of the problems that women encounter. The merit of this project is that once armed with this insight, chemists and other scientists and engineers can be better equipped to encourage, promote, and institute practices and procedures that are supportive of women (and men) in academe.

The Social Science Perspective

The book begins with a discussion of possible reasons for the low participation of women in science; this first section, from the perspective of social scientists, confronts numerous examples of the barriers that women must overcome in order to be successful academicians. Virginia Valian is a Professor of Psychology and Linguistics at Hunter College in New York City, as well as a Professor at the Graduate Center of the City University of New York (CUNY). She is the Principal Investigator on an NSF ADVANCE Institutional Transformation Award at Hunter College, and is the author of the compelling book, "Why So Slow? The Advancement of Women." In her opening chapter, Dr. Valian sets the general tone for our book by addressing the reasons that for the continuing lag women have behind men in the sciences. She also outlines measures and practices that institutions can use to eliminate gender disparities and help all scientists flourish.

Next, Mary Frank Fox focuses more specifically on the gender disparities in academe by discussing the findings from her analyses of salaries, publications, productivity, work attitudes and behavior, and educational and career patterns among scientists and academics. Dr. Fox, also a social scientist, is an NSF ADVANCE Professor of Sociology in the School of History, Technology, and Society at Georgia Institute of Technology, as well as the Co-Director of the Center for Study of Women, Science and Technology. She has published numerous times on the gender disparities experienced by faculty in scientific departments. Her unique expertise in gender, science, and academia is well suited to this discussion.

Another social scientist weighing in on the issue of gender disparity is Susan A. Nolan, an Assistant Professor of Psychology and the Director of the Center for Women's Studies at Seton Hall University in South Orange, New Jersey. Dr. Nolan's chapter rounds out the Social Science Perspective on the status of female chemists by describing the gender disparities specific to the training and mentoring of academic chemists. Her insight is founded upon data garnered from a comprehensive, collaborative research project conducted by the editors of this book (two psychologists and two chemists) titled, "The Career Continuity Survey," which was generously supported by the Camille and Henry Dreyfus Foundation (SG-02-072) (*1, 3, 4, 11*). Based upon the findings from this recent survey of male and female doctoral students graduating from top ranked universities in chemistry, Dr. Nolan elaborates upon many of the obstacles that academic women in scientific disciplines perceive that they face—and indeed have experienced. She centers much of the career challenges and difficulties faced by female chemists within the context of mentoring relationships.

Taken together, these pieces document how women are not considered to be full partners in chemical education. Prior (mis)perceptions of women's abilities in the form of negative stereotypes projected by others may interact with negative self-perceptions in the form of low esteem, under-developed scientific identities, or diminished self-efficacy that are reinforced in training and work environments. Together, these factors may interact with women's subsequent training experiences to impact negatively women's participation in chemistry (*2, 10, 12, 14, 15*).

Recent Findings from Survey Data

The next segment of the book more specifically focuses upon findings from recent survey studies of female and male faculty members and doctoral recipients who have sought academic employment. First, Cecilia H. Marzabadi, an Assistant Professor of Chemistry and Biochemistry, and a faculty member in the Center for Women's Studies at Seton Hall University supplements the discussion of data from the same survey study introduced by Dr. Nolan. Instead of presenting data on mentoring relationships, however, Dr. Marzabadi highlights patterns that have emerged in the specific employment paths of survey respondents. In particular, Dr. Marzabadi traces the academic development of women and men in chemistry through graduate school and post-doctoral training into their initial employment positions. She presents data that identify reasons for the different career choices made by men and women, and suggests that these reasons lead to a smaller proportion of women receiving "qualifying" requisite experience for prestigious academic jobs. She also cautions that rather than simply explaining the paucity of female academicians with the "usual" female flight scenario (women choosing to leave, even fleeing, for family reasons), we must acknowledge that the underlying cause for the under-representation of women in top-ranked chemistry departments may instead be the direct result of mentoring and hiring practices that give priority to male candidates (e.g., more encouragement for men to take prestigious post-doctoral positions, which better "qualify" them for employment at top-ranked departments). Such factors build upon each other and lead to disparate career paths for men and women in academic departments in STEM fields, most specifically in chemistry.

Sue V. Rosser also presents survey data that speak to gender disparities in academe. Dr. Rosser is Dean of the Ivan Allen College, the Liberal Arts College at Georgia Institute of Technology. She has authored the book "Re-Engineering Female Friendly Science." In her presentation, Dr. Rosser highlights her research on the theoretical and applied problems of women in science; in particular, she attends to the institutional barriers for female scientists and engineers by drawing upon data from NSF POWRE and Clare Boothe Luce Professorships. Moreover, like the data from our own Career Continuity Study, Dr. Rosser's research points to the perpetuation of un-inviting or uncomfortable work environments that induce women to make choices leading them away from academic careers, or even avoiding this career path altogether.

Promotion and Retention Issues

One particularly positive outcome of research on disparity is the development of effective programs to assist in the training, mentoring, and support of women who actively seek prestigious academic positions. Jane Z. Daniels presents one such model of effective support for junior women in academic positions. Dr. Daniels reports on ways that the Clare Boothe Luce Program has impacted positively the careers of tenure-track women, particularly those at the Assistant or Associate Professor level. She is Director for the Clare Boothe Luce Program, the most significant source of private support for women in science, engineering, and mathematics. Many women have made great strides in their scientific careers due to the resources afforded them by the program.

Anne Preston, an economist from Haverford College and the author of the compelling book "Leaving Science. Occupational Exit from Scientific Careers", also contributes a paper on the issue of the retention of women scientists. Using information garnered from public databases and from survey data, Dr. Preston gives estimates on the numbers of female scientists leaving the workforce, examines the factors contributing to their exit and makes recommendations for retaining women in the scientific careers.

The Status of Women Chemists in Europe

Also presented are the efforts female chemists in Europe are making in obtaining academic positions. Their success has far reaching implications, and gathers a global view of the progress that female chemists are making. For instance, women in Sweden have a significantly higher employment rate in academe as compared to other countries (*16*). By highlighting the successful practices and policies used in some countries, we may assist institutional reform in other countries, including the United States. This broader impact is compelling. By linking efforts in the U.S. and abroad, we will be better able to understand the barriers that must be overcome to increase substantially the numbers of women holding tenure-track positions on chemistry faculties at Ph.D. granting institutions across the globe.

To this end, the status of female chemists beyond the borders of the United States is examined in a very insightful discussion in which the similarities in inequities that women face in the United States and Germany are considered. The scientist contributing to this section posits that several behavioral patterns are likely common to the two nations (e.g., the implementation of unwritten rules of standards). Sonja Schwarzl contributes a very interesting perspective on the status of female chemists by attending specifically to the state of academe in Germany. In her paper, she presents recent data on the success rate for women in obtaining doctorates in chemistry and the gender composition of the chemistry faculties in Germany. One particularly striking finding related by Sonja Schwarzl is that Germany once produced an equivalent number of advanced degrees in chemistry to the United States, but is now experiencing a severe drought in both male and female chemistry students. Sonja Schwarzl is a doctoral student in chemistry at the University of Heidelberg, Germany, and also is a member of the executive committee of the Society of German Chemists' Working Group for Equal Opportunities in Chemistry.

Hiring Practices

Valerie J. Kuck, a Visiting Professor in the Department of Chemistry and Biochemistry and in the Center for Women's Studies at Seton Hall University in South Orange, New Jersey contributes a discussion of the hiring practices in academic STEM fields. Recently retired from Lucent Technologies, Bell Laboratories, Valerie Kuck has a broad understanding of the issues that women face at all levels in the field of chemistry. Her research on the doctoral attainment rates of men and women at the top universities in chemistry has evoked much debate by chemists. More specifically, data collated by our collaborative research group show preferential hiring of doctoral graduates from a select number of elite universities, with some schools becoming the "top suppliers" of new faculty to "top" departments.

In her chapter, she summarizes much of this work and discusses the hiring patterns at the top fifty ranked universities in chemistry. In particular, she compares the academic employment of male and female doctoral graduates from the same university (a "yield" and "parity" index), and computes an "impact factor," by which universities can be rated for their ability to place their graduates on faculties at the top fifty universities.

On a positive note, recent data show that in comparison to older faculty, younger faculty members are coming from a broader range of institutions In the present chapter, Valerie Kuck tracks the representation of women in the baccalaureate pool, entering graduate school, and compares female doctorate and post-doctoral position attainment rates to those of their male peers, She also identifies the sub-fields women choose to study and the difficulty that women in organic chemistry are having in obtaining academic positions. Finally, she shows that women are represented at higher levels on the faculties granting the lesser terminal degrees. With respect to gender patterns in these hiring practices, it appears that women still are confronting barriers to equal employment, especially at the PhD-granting institutions.

Conclusion

In recent years, chemists have become more aware of the low representation of women in tenure-track positions on the faculties at Ph.D.-granting institutions. For example, women only comprised about 11% of chemistry faculties at the top fifty-federally funded departments in 2002 (*3*). This growing awareness is largely the result of the publication of a number of articles in the chemical trade literature that describe this situation. These articles present the gender composition at Ph.D.-granting universities; however, there have been few detailed analyses of the factors influencing the under-employment and low retention of women in academe (*7*). Women are more represented at the student level in chemistry; for example, 50% of the B.S. degrees in chemistry were awarded to women in 2002, and more than a third of the graduate students in this field are women (5). As already mentioned, practices must be developed in order to cultivate and maintain a body of faculty members that are representative of the demographic features of the individuals whom they mentor. In order to develop and to put into effect such practices and policies, an understanding of the career issues faced by women—from students to administrators—must be tendered.

This symposium series book highlights the research of scientists who have studied the difficulties encountered by female academics in the physical sciences and engineering. Each chapter brings a unique offering to the discussion table. With this knowledge, chemists will have a far better understanding of the problems that women face, and they will be better equipped with tools to encourage, promote, and institute practices supportive of women and men in academe. An enhanced understanding by chemists of the barriers that must be overcome to increase substantially the numbers of women on chemistry faculties at Ph.D.-granting institutions could in fact serve as a model to prompt more wide-spread institutional reform. Indeed, whereas much of the discussion in the original ACS symposium and this ensuing collection of papers are centered upon the identification of barriers to equalizing the success rates of women and men in science, we also must point readers to the good efforts that are employed in hiring, promotion, and retention of women in academe. These efforts are elaborated as "best practices" utilized by some institutions, and deserve thoughtful consideration.

In short, the major contribution of this book is that it delineates clearly the ongoing and international struggle for equality experienced by women in academe. Moreover, it is strikingly apparent that the obstacles faced by women are not unique only to chemistry, but to other STEM and social science fields as well. Furthermore, though our discussion is specific to the female gender, it is not true that such issues are limited to women only; the obstacles encountered in academic training, hiring, promotion, and tenure of individuals extends to other minority groups as well. Further, the problems with retention in the field are germane to women and men alike. Departmental and institutional dialogues about the implications inherent in the following chapters are not only useful, but even necessary where commitment to dissolving disparity is valued and integral to productive research, and mentoring.

References

1. Nolan, S. A.; Buckner, J. B.; Marzabadi, C.H.; Kuck, V. J. Manuscript submitted to *Sex Roles.*
2. Kuck, V. J. Refuting the leaky pipeline hypothesis. *Chemical & Engineering News.* **2001,** *79,* pp. 71-73.

3. Kuck, V. J.; Marzabadi, C. H.; Nolan, S. A.; Buckner, J. P. Analysis by gender of the doctoral and post-doctoral institutions of faculty members at the top fifty ranked Chemistry departments. *J. Chem. Ed.* **2004**, *81*, pp. 356-363.
4. Marzabadi, C. H.; Kuck, V. J.; Nolan, S. A.; Buckner, J. P. Career outcomes of doctoral graduates from top-ranked chemistry departments: Results from a career continuity study. *Unpublished manuscript.*
5. National Science Foundation, Division of Science Resources Statistics. *Science and engineering doctorate awards: 2002. Arlington, VA: National Science Foundation.*
6. Fox, M. F. In *Equal rites, unequal outcomes: Women in American research universities:* Hornig, L.S., ED. New York: Kluwer Academic/Plenum Publishers: New York, NY, 2003; pp. 91-109.
7. Preston, A. E. *Leaving science.* Russell Sage Foundation: New York, 2004.
8. Long, J. S. *From Scarcity to Visibility: Gender Differences in the Careers of Doctoral Scientists and Engineers*; National Academy Press: Washington, D.C., 2001.
9. Sharpe, N. R.; Sonnert, G. Women mathematics faculty: Recent trends in academic rank and institutional representation. *Journal of Women and Minorities in Science and Engineering.* **1999**, *5*, pp. 207-217.
10. Valian, V. *Why So Slow? The Advancement of Women.* The MIT Press: Cambridge, MA, 2000.
11. Buckner, J. P.; Kuck, V. J.; Marzabadi, C. H.; Nolan, S. A. *Examining the role of undergraduate training in the academic career path: Does gender matter?* Paper presented at the 229th National Meeting of the American Chemical Society, San Francisco, CA, **2005**.
12. Etzkowitz, H.; Kemelgor, C.; Uzzi, B. *Athena Unbound, The Advancement of Women in Science and Technology*; Cambridge University Press: Cambridge, U.K., 2000, Chapter 6, 83-103.
13. *NSF-NIH Survey of Graduate Students & Post-doctorates in S&E.*
14. Rosser, S. V.; Zieseniss, M. Career issues and laboratory climates: Different challenges and opportunities for women engineers and scientists (Survey of Fiscal Year 1997 Powre Awardees). *Journal of Women and Minorities in Science and Engineering,* **2000**, *6,* pp. 95-114.
15. Schlegel, M. Women mentoring women. *American Psychological Association Monitor,* **2000**, *31,* pp. 33-36.
16. Olsson, Y. *Swedish women in science: Towards real equity?* Paper presented at the 226th National Meeting of the American Chemical Society, New York, 2003..

Chapter 2

Women, Science, and Academia

Virginia Valian

Department of Psychology, Hunter College and CUNY Graduate Center, 695 Park Avenue, New York, NY 10021

Women are underpaid and underpromoted across the professions generally and in academic science in particular. For relevant references, see the other chapters in this book, (*1, 2, 3*), the National Science Foundation's (NSF's) periodic reports, and American Association of University Professors' (AAUP's) annual data. Helpful websites include the Gender Equity Project at Hunter College: www.hunter.cuny.edu/genderequity and the gender tutorials at www.hunter.cuny.edu/gendertutorial. data. I will not review the data here but concentrate on how to explain the data.

The data support the following generalization: women and men are close to equality in salary and rank when they begin their professional lives – in every field – but women's advancement lags behind men's. The data also support the conclusion that the problem is general, occurring in all the professions - business, medicine, law, and academia. To me, that signals the need for a general, social-cognitive explanation, an explanation that will cover the ubiquity of the phenomena. My explanation will use two key concepts: gender schemas and the accumulation of advantage.

Problems of gender equity in academia affect faculty and students. Faculty (male and female alike) do not fully understand how gender works: that produces less effective teaching and mentoring of students, especially female students, and less effective management, for women, of their own careers. Even when faculty and students are aware of gender problems, they lack an educated understanding of the reasons that women are underrepresented in leadership positions in the professions.

Compounding the problem is the fact that neither students nor faculty are well versed in techniques and strategies for changing their institution or their discipline. There are few mechanisms whereby women and men can learn how to improve the procedures that determine who is appointed as a leader, who is awarded an honor, who is highly recommended for graduate and post-doctoral study, and so on. Nor are there opportunities for women to understand why they themselves would be reluctant to acknowledge their status as members of an underrepresented and undervalued group.

An improved educational environment for young people – and an improved working environment for everyone – requires knowledge not just of the facts on men's and women's professional advancement but, more importantly, an understanding of the social psychological and cognitive mechanisms that underlie men's advantage and women's disadvantage. To develop successful strategies for improving the educational and professional environments in which they learn and work, students and faculty need more than anecdotes and impressions. They need a databased approach, which will provide an analysis from which they can work for change with others.

Let me turn, then, to explaining what schemas are. They are largely nonconscious hypotheses about the characteristics of social groups. The term "schema" is similar to the term "stereotype", but I prefer schemas because it is more neutral and brings out the proto-scientific nature of schemas. Schemas are partially diagnostic. They help us to predict people's behavior and are thus useful to us. They also, however, can lead us astray. Gender schemas assign different personality characteristics to men and women. We see men as capable of independent action (agentic), doing things for a reason (instrumental), and as getting down to the business at hand (task-oriented). We seem women as taking care of others (nurturant), showing concern about others' welfare (communal), and have articulating their feelings (expressive).

There is evidence that, indeed, men have more of the traits we think of as characteristic of them, and that women have more of the traits we associate with them (*4,5*). Although it is possible to have all the traits, some of each, or none, we tend to think in terms of mutual exclusivity. We thus tend not to recognize men's ability to care for children and tend not to appreciate women's professional competence. Not only do we tend to underestimate people's capacities, we also tend to underestimate their interest in the fields that seem less compatible with the dominant traits we associate with them.

Gender schemas are important in judgments of competence, ability, and worth. In professional settings, they result in our overvaluing men and undervaluing women. I say "our" because men and women do not differ in their subtle evaluations. Both groups overrate men and underrate women. A broad array of sources confirms that conclusion. Here are two recent examples.

The first, by Heilman and her colleagues (*6*) investigates how males and females rate people who are described as being an Assistant Vice President in an aircraft company. For this experiment, the researchers deliberately chose a male-dominated field. The evaluators read background information about the person, the job, and the company. In half the cases, the assistant VP is described as about to have a performance review. In that case, the evaluators can't tell how well the person is doing in the job and have to make a guess. In the other half of the cases, the person is described as having been a stellar performer. No matter what materials the evaluators received, their job was the same: rate how competent the employees are and how likeable they are.

When no information is provided about how well people are doing in the job, evaluators rate the man as more competent than the woman, and rate them as equally likeable. When the background information makes clear that the woman is extremely competent, evaluators do rate the man and the woman as equally competent, but they rate the woman as much less likeable than the man. They also perceive the woman as considerably more hostile than the man.

Thus, in evaluating a woman in a male-dominated field, observers see her as less competent than a similarly described man – unless there is clear information that she is competent. And in that case, they see her as less likeable than a comparable man. Notably, as is the case in almost all such experiments, there were no differences between male and female subjects. Both males and females see competence as the norm for men and as something that has to be demonstrated unequivocally for women. Both males and females see competent men as likeable. Neither males nor females see competent women as likeable.

And likeability matters: in a follow-up experiment, the experimenters described targets as high or low in competence and high or low in likeability. People rated the targets who were high in likeability as better candidates for being placed on a fast track and as better candidates for a highly prestigious upper-level position. We cannot tell women just to be competent, because what can make the difference in some rewards is likeability. Again, there were no male-female differences.

Many of the situations where we take men more seriously than women are small ones. The differences in the experiment just described are not enormous, though they are reliable and significant. We are tempted to dismiss small differences as unimportant. In a meeting, for example, we might ignore a woman's idea but pay attention to and praise that same idea when a man produces it 10 minutes later. Such small events seem like molehills that women should ignore. But a computer simulation demonstrates that a tiny advantage in favor of men – an advantage that accounts for only 1 % of the variability in outcomes – results over the long haul in a substantial advantage, as a computer simulation by Martell, Lane, and Emrich (*7*) demonstrates. Mountains *are* molehills, piled one on top of the other. It is like interest on an investment. Even if your advantage is small, if it is systematic and occurs repeatedly, it mounts up to create a mountain of difference.

The second study, by Norton and his colleagues (*8*) demonstrates how people shift their standards in order to justify a choice that seems a priori reasonable to them. In this experiment, gender schemas determined what seemed reasonable. The experiments asked male undergraduates to select a candidate for a job that required both a strong engineering background and experience in the construction industry. Again, then, the experimenters chose an occupation that most people would see as more appropriate for men than for women.

The evaluators rated 5 people, only 2 of whose resumes were competitive. One candidate had more education – both an engineering degree and certification from a concrete masonry association – than the other, who only had an engineering degree. The other candidate had more experience – 9 years – than the other, who only had 5 years. In the control condition, only initials identified the candidates. Here, the evaluators chose the candidate with more education three-quarters of the time and about half rated education as the most important determinant of their decision. Thus, the undergraduates saw education as more important than experience.

In one of the experimental conditions, a male name was given to the resume that had more education and a female name to the resume that had more education. Again, evaluators chose the candidate with more education three-quarters of the time and about half rated education as very important. In the second experimental condition, a *female* name was given to the resume with more education and a *male* name to the resume with more experience. If people were unaffected by gender schemas, they would again pick the person with more education even though that person was female. But that was not what happened. Instead, less than half the evaluators picked the person with more education and less than a quarter said that education was the most important characteristic.

Men look more appropriate than women for the job of construction engineer, whether they have more education or more experience. The standards by which we judge people will shift depending on our a priori judgments about their goodness of fit. Gender schemas help determine goodness of fit. The implications of this experiment for hiring in academia are plain. We are at risk of giving a man credit for being a man and downgrading a woman with superior credentials because she is a woman.

Implications for Remedies

Let's turn now to the implications of research on schemas to considering how we will change things. We will not change our schemas any time soon. Schemas resist change. There has been change and there will continue to be change, but I chose experiments published in 2004 in order to demonstrate that schemas are alive, well, and active.

Our first task, in my view, is to understand how schemas work, their persistence and ubiquity, and the limitations they set on our ability to judge others accurately. Know the data, know the theory; communicate the data, communicate the theory. Use the data and the theory to inform proposals for remedies. I offer a few here.

Remedy 1. Since we cannot rely on our evaluations, we need policies and procedures that will allow us to check our work and correct it if necessary. Such procedures will constitute a new set of operating minimal norms. For example, whenever we create a list of colloquium speakers, or a slate of candidates, we will check the list to see if it matches the percentage of women in the pool. If it doesn't, we will search harder for qualified candidates. This is the way we avoid errors.

New operating norms can be introduced. For example, I recycle my newspapers in the blue container outside my back door, not because I am strongly committed to ecology, and not because anyone will put me in jail if I don't, but because it's part of being a reasonable citizen and because it's easy to do. We need to develop equally simple procedures for equity and diversity.

Remedy 2. Since schemas are persistent and ubiquitous, we need distributed leadership, that is, leadership not only from those in official leadership positions but leadership from those who are working members of their institutions. Students – undergraduate and graduate – can be effective, post-docs and faculty can be effective. It's our institution. Only we can make it work. Most all of us will need some training, since few of us know how to develop and carry out effective strategies. The NSF ADVANCE Institutional Transformation Awards and Leadership Awards are an example of creating leadership opportunities – and, very importantly, resources – for women in academic science. One result of the awards is a wide array of creative initiatives. We will be able to learn from these initiatives what works, what doesn't, and why. In addition, now that 19 schools have Institutional Transformation Awards, we are in a position to act as a group to affect policy.

Remedy 3. We can learn from the leadership literature about how to be more effective, regardless of our locus within the institution, how to create leaders, and how to be effective in improving the value and effectiveness of our institutions. Many characteristics of effective leaders are within our grasp, regardless of our locus: we can articulate our goal – gender equity; we can supply reasons for people to see that goal as desirable; we can develop specific recommendations strategies for how to get to our goal; we can build alliances. Finally, since we all have some sphere of influence, we can implement our plans.

References

1. Valian, V. *Why so slow? The advancement of women.;* MIT Press: Cambridge, MA; 1998.
2. *From scarcity to visibility;* Long, J.S., Ed.; National Academy Press: Washington, D. C., 2001.
3. Xie, Y.; Shauman, K. A. *Women in science: Career processes and outcomes;* Harvard University Press: Cambridge, MA, 2003.
4. Spence, J. T. & Helmreich, R. L. *Masculinity and femininity: Their psychological dimensions, correlates, and antecedents;* University of Texas Press: Austin, TX, 1978.
5. Spence, J. T. & Sawin, L. L. *Women, gender, and social psychology;* Images of masculinity and femininity: A reconceptualization. In V.E. O'Leary; R.K. Unger; B.S. Wallston, Eds.; Erlbaum: Hillsdale, NJ, 1985; pp 35-66.
6. Heilman, M. E.; Wallen, A. S.; Fuchs, D.; Tamkins, M. M. *J of Applied Psychology.* **2004,** *89, 416-27.*
7. Martell, R. F.; Lane, D. M.; Emrich, C. *American Psychologist.* **1996,** *51, 157-158.*
8. Norton, M. I.; Vandello, J. A.; Darley, J. M. *Journal of Personality and Social Psychology.* **2004,** *87, 817-831.*

Chapter 3

Women and Academic Science: Gender, Status, and Careers

Mary Frank Fox

NSF Advance Professor, School of Public Policy, Georgia Institute of Technology, Atlanta, GA 30332–0345 (phone: 404–894–1818, fax: 404–371–8811, email: mary.fox@pubpolicy.gatech.edu)

This paper addresses sources of women's depressed status in academic science, and proposes solutions for women's advancement. It does so by analyzing the role of individual characteristics, and the role of organizational features of doctoral education and of academic work, in explaining the career attainments of women in science; as well as the role of these factors in solutions for women's advancement in academic science.

Scientists work in numbers of employment sectors—in education, in private-for-profit and not-for-profit organizations and industries, as well as in federal, state, and local government, and self-employment. However, as of 2001, the majority (52%) of employed women who had doctoral degrees in scientific fields[i] were working in educational institutions, and the vast (90%) preponderance of these doctoral-level women in education were in four year colleges or universities—academia (*1*).[ii]

Women in academic science are a critical group for analysis of gender, status, and careers. Women in academic science have already survived series of barriers of selection—both self-selection into science fields and selection by institutions. They have moved through the proverbial pipe-line. They have completed doctoral degrees, and they have the credentials for professional work. However, the highest career attainments often elude this select group of women. As of 2001, in mathematical and physical sciences, women were still less than ten percent, and in engineering less than five percent, of the full professors in universities and four year colleges. In computer and information sciences, in life sciences, and even social sciences, women were less than twenty percent of the full professors (Table I).

Table I. Doctoral Scientists and Engineers Employed in Four-year Colleges and Universities, by Field and Rank, 2001

Field	*Total**	*Full Professor*	*Associate Professor*	*Assistant Professor*
Physical Sciences	37,140	13,760	6,600	5,710
%Women	*15.5*	*6.4*	*16.5*	*24.9*
Mathematical Sciences	14,980	6,720	3,740	2,480
%Women	*14.5*	*8.7*	*15.8*	*25.8*
Computer Specialties	3,760	770	1,710	920
%Women	*21.8*	*16.6*	*20.1*	*25.6*
Biological/life Sciences	72,850	22,330	13,740	14,140
%Women	*30.3*	*16.4*	*29.0*	*35.9*
Psychology	30,190	9,710	5,960	6,530
%Women	*48.2*	*28.6*	*47.3*	*61.6*
Social Sciences	47,240	18,980	11,680	9,470
%Women	*31.7*	*19.5*	*34.9*	*43.0*
Engineering	27,110	11,270	6,140	4,900
%Women	*8.4*	*2.8*	*9.4*	*14.9*
Total, all fields**	245,060	86,400	52,920	47,790
%Women	*28.4*	*15.5*	*29.6*	*38.9*

*Total includes instructor/lecturer, other faculty, "does not apply," and "no report."
**Total fields include Health Sciences.

SOURCE: Commission on Professionals in Science and Technology. *Professional Women and Minorities: A Total Human Resource Data Compendium* [15th edition]. Washington, D.C., 2004: Table 3-46.

These relatively low proportions of women at full professorial rank exist despite the rates of increase in the number and proportions of women earning doctoral degrees in scientific fields over the past three decades, and the passage of years for women to mature in professional time and experience. In life sciences, the proportion of women among doctoral recipients increased from 12% to 18% between the 1960s and 1970s; by the 1980s, women were earning 29%, and in the first half of the 1990s, 36% of the doctoral degrees in these fields. In the mathematical, physical, earth, and atmospheric sciences, the proportions of women among doctoral recipients are lower, but across these fields[iii], women were earning 8% of the doctoral degrees in the 1970s, 15% in the 1980s, and 20% in the first half of the 1990s (*1*).

This raises questions about the sources of women's depressed status in academic science, and solutions for their advancement. What accounts for women's status in academic careers? What is the role of individual characteristics in explaining women's career attainments? What is the role of organizational features of doctoral education and of academic work and the workplace? What are the implications for supporting women's advancement in academic sciences?

Accounting for the Status of Women in Academic Science

Individual Characteristics

In explaining career outcomes in scientific fields, personal/individual factors play a part. But such individual characteristics do not exist in a social vacuum, and by themselves, do not explain educational and career outcomes in science. For example, social and attitudinal factors have been shown to have a stronger effect upon high school students' enrollment and grades in mathematics than do variations in measures of mathematical aptitude (*2*). Further, no direct relationship has been found between measured creative ability or intelligence and outcomes of productivity, among those in scientific fields (*3, 4*). Rather, organizational conditions in the workplace, such as autonomy in a decentralized atmosphere (5) or a pool of resources in excess of minimum needs (6), are important. The presence or absence of these conditions may enhance or block the translation of people's creative characteristics into productive or innovative "outputs" (*5, 6*). In addition, although women's career attainments are lower than men's, their measured ability (IQ) is higher. Data on IQ may not be adequate indicators of intelligence or ability; but to the extent that they capture differences, they indicate that, if anything, women in scientific fields are a more select intellectual group than men (*7*).

One might ask, then, are women and men receiving degrees from different types of institutions? Are women's doctoral degrees incomparable to men's, that is, from lower-ranking institutions? The answer is no. Women and men are about as apt to have doctoral degrees from top-ranking institutions. Across fields, the pattern is one of similarity in the prestige of doctoral origins of women and men (*8*). It is not simply a matter then of boosting the doctoral origins of women.

Likewise, gender differences are small in certain indicators of financial support for graduate training, measured as percentages of women compared to men who held research or teaching assistantships (*9*). However, these data do not specify the quality or character of assistantships or training (*10*). Certain clues to educational and career outcomes lie in these factors, which I address in the following section.

Household and family statuses represent another set of individual characteristics. The mythology of science (*11*) has it that good scientists/academics are either men with wives or women without partners or children. Yet, the evidence contradicts this conventional wisdom. Data indicate that while marriage negatively affects rank and salary of academic women, the effects are not significant except in the case of salary within research universities (*12*). Among biochemists, marriage had a positive effect on being promoted from assistant to associate professor for both women and men; and for promotion to the level of full professor, marriage had no effect (*13*).

Further, married women publish as much women who are not married. This has been found across physical, biological and social sciences (*14, 15, 16, 17*). Moreover, among various samples of academic scientists, children had either no effect on women's productivity (*15*); a slightly, negative, nonsignificant effect (*18, 19*); or a positive effect (*14, 16, 20*).

However, these data describe only those women who <u>hold</u> academic positions, and have survived a demanding process of scrutiny, selection, and evaluation. Household and family demands may take their toll along the way, so that women with children are less likely than men to continue education in science, and if they do attain advanced degrees, are less likely to pursue careers in science and engineering (*21*). Thus, women who "fall out" of scientific education and professional participation do not appear in the data bases of academic scientists in the studies reported in the previous paragraph. Marriage and young children can have a multitude of effects in personal sacrifices as well as rewards, and extraordinary arrangement of accommodation (*22*). What the data show is that marriage and young children are not associated with depressed productivity of women who do have academic appointments in science.

Organizational and Environmental Factors

In understanding the status and performance of women in academic science, we need to look also to features of the organizations in which people study and work. Women's status in academic science is not a simple reflection of their background, ability, and skills. Rather, it is a consequence more so of complex factors of organizational context—the characteristics and practices of the settings in which people are educated and work.

Organizational settings are important to the attainments of women—and men—across occupations. But they are especially important in scientific fields. This is because scientific work revolves on the cooperation of people and groups; and requires human and material resources. Compared to non-science fields, the sciences are more likely to be conducted as teamwork rather than solo; to be carried out with costly equipment; to require funding; in short, to be interdependent enterprises (*23, 24*).

Social and Organizational Features of Doctoral Education

In a study of social and organizational features of doctoral education in science and engineering fields, I addressed characteristics and practices of departments, research teams, and advisement in doctoral education in departments that had been high, low, or improved in doctoral degrees awarded to women over a 17 year period (*25, 26, 27, 28, 29*). The study includes a survey of 3,300 students in sixty one doctoral-granting departments of chemistry, computer science, electrical engineering, and physics, conducted in 1993-94.[iv] The response rate of students across fields was 61%.

The survey of students points to different experiences and outcomes for women and men students in departments, research groups, and with advisors. For example:

1. In experiences <u>within their departments</u>, women are less likely than men to report that they are taken seriously by faculty, and that they are respected by faculty.
2. In their experiences <u>in research groups</u>, compared to men, women report that they are less comfortable speaking in group meetings. Despite strong preferences for collaboration among both men and women students, women report collaborating with fewer men graduate students and men faculty members in the preceding three years.
3. In <u>adviser-advisee arrangements</u>, compared to men, women are less likely to report that they have received help from advisers in crucial areas, such as learning to design research, to write grant proposals, to coauthor

publications, and to organize people. Women are also more likely than men to report that they view their relationship with their adviser as one of "student-and-faculty" compared with "mentor-mentee" or as "colleagues," which may suggest greater social distance for women students.

These findings suggest different opportunities for women and men to participate in research groups, to collaborate, and to gain significant roles in the scientific enterprise. Such matters of gender, social context, and participation are important because, as discussed, academic science is a social system--of communication, interaction, and exchange. If women are more constrained within the social networks of academic science—in departments or beyond in disciplinary communities—this has consequences for status and performance.

Social and Organizational Features of Work and the Workplace

Social and organizational features of work and the workplace, following graduate school, are consequential in understanding career attainments. More so than men, women are outside of the networks in which human and material resources circulate. For example, a survey of all women faculty and a stratified random sample of men faculty in four Colleges (Computing, Engineering, Sciences, and Liberal Arts) at Georgia Institute of Technology (Georgia Tech), a leading scientific and technological institution, points to gender differences in departmental work environments. Women report less frequent interaction around research with faculty in their units; they give lower ranking than men to equipment available to them; and they characterize their units as less "helpful" than do the men (*30*).

What about collaboration, specifically, as an explanation of women's career attainments? The evidence and answer are mixed. Studies of sociologists have reported that women are both more likely (*31*) and less likely (*32*) to collaborate. In an analysis of a matched sample of women and men who received their Ph.D.s in six scientific fields, women were as likely as men to coauthor papers (*15*); this is corroborated among a sample of biochemists as well (*33*). The issue may be more subtle, however, than simply rates of co-authorship. Women may have more difficultly finding and establishing collaborators and may have fewer collaborators available to them (*33, 34*). In keeping with this, a survey of faculty at Georgia Tech indicates that women and men are as likely to report that they have colleagues in their home unit who work in a research area related to their own; women, however, are less likely to report the "willingness" of these faculty to collaborate with them (*30*).

This leads to consideration of publication productivity. In the analysis of gender and career attainments, publication productivity is important for two reasons. First, publication is the central social process of science, because it is through publication that research findings are communicated and verified, and that priority of work is established (*35, 36, 37*). Second and accordingly, until we understand productivity differences, we cannot adequately address other gender differences in location, rank, and rewards, because they are related to—but not wholly explained by—publication.

Although the gender gap in publications has been narrowing recently in biological and social sciences, women publish less than men, especially in physical sciences (*38, 39, 40, 41*). Women's depressed publication productivity is both cause and effect of their career attainments. That is, it both reflects women's location in lower ranks and lower ranking institutional locations, and it partially accounts for it (*42*). "Partially" is a key term: holding constant levels of publication productivity, women's career attainments, particularly academic rank, remain lower than men's. Although understanding is incomplete of the underlying processes, women are promoted at lower and slower rates, after controlling for numbers of articles published and citations to articles (*7, 13, 39*). This also holds among different types of institutions as they vary by prestige level.

This brings us, in turn, to evaluative schemes. In the relationship between social and organizational environment and career attainments, the process of evaluation has consequences for rank and rewards. Key here is that in academic science, performance may be judged against a standard of absolute excellence; and that, in turn, becomes a subjective assessment (*23*). Experimental data indicate that when standards are subjective and loosely defined, it is more likely that persons with majority status characteristics are perceived to be the superior candidates and that gender and racial/ethnic bias may operate (*39*). These evaluative processes may help explain career attainments as well.

Implications for Advancement of Women in Academic Science

In the assessment of women and status in academic careers, we need to get beyond issues of numbers of women. Since the 1970s, women have increased in number and proportion of doctoral degrees awarded. But numbers do not ensure significant participation and performance, and numbers of women with doctoral degrees do not necessarily change patterns of gender, status, and hierarchy in science. Women have long been present in science, though not in valued, highly rewarded, or even visible roles (*43, 44*). Further, for science as for other professions, the relationship between gender, education, and status is complex. It is not a simple matter of increasing education and increasing social and economic status. Women's educational attainments do not automatically translate into career-attainments, especially advancement in rank, on a par with men's (*39, 45*).

Among the factors that explain women's lower career success, individual characteristics of ability and marriage/motherhood account modestly or little. Likewise, prestige of doctoral origins does not provide an explanation. Because women and men have different experiences in graduate training, better clues lie in the nature and patterns of advising, collaboration, and apprenticeship in doctoral education, as discussed. In like manner, the career attainments of women are conditioned by aspects of the social and organizational environments of work and the workplace.

By implication, this means that improvement in women's attainments in careers—their status and performance—will not depend merely upon the detection, cultivation, and enhancement of individuals' ability, skill, and background. Improvement in women's advancement means attention to organizational and environmental factors such as allocation of resources, access to interaction and collaboration, and the operation of evaluative schemes–in departments, colleges, and institutions of academe.

Further, environments do not operate uniformly for people and groups. Different environments for women and men are not simply a matter of women and men being located in different types of institutions or settings–large compared to small, more compared to less resourceful, or more compared to less prestigious. Rather, within the same type of setting–indeed, the identical setting–women can have fewer (and different) experiences with faculty, collaborative arrangements, and enabling help–with consequences for participation, performance, and career attainments (*23, 30*).

In addition, merely increasing the numbers of women may not alter "norms" or "standard practices" of education and work—with implications for gender and status. To illustrate: research by Sonnert and Holton (*41*) shows that women in science exercise more care, caution, and attention to detail in their publications; and they are more likely to confirm and integrate findings before releasing them for publication. Therefore, if women have certain approaches, as in tendencies to confirm findings before publishing, and if they need to conceal, obscure, or even "overcome" such approaches, then a higher number of more fragmented pieces of published work may continue to constitute an unchallenged standard for scientific productivity (*42, 46*). This can prevail even though numbers of women in science increase.

However, certain organizational practices and policies can be shaped to support more equitable advancement of women in academia:

1. Written guidelines and specified benchmarks for evaluation and promotion support equity, while "flexible" and informal processes of evaluation tend to support those who "look like" those currently in power (*23, 39*).

2. Relatedly, start-up funds for laboratories, teaching assignments, and release time from teaching that reflect allocations according to clear and understood standards, rather variable "administrative favors," support greater equity in experiences and outcomes in academic science.
3. Practices of incorporating junior faculty into collaborative networks in the department affect research opportunities in highly collaborative fields (*47, 48*).
4. The publication of more fragmented pieces of publication can be discouraged by lowering the rewards and incentives for the practice, as federal agencies have done, by limiting in proposals for grants the listings of applicants' five most important publications.

The point here is that just as organizations are structured, so they may be restructured (*27*). In academic science, this means examining existing ways of organizing departments, evaluating faculty, and distributing resources, toward support of improved equity in research performance, collegial opportunities, and advancement of all faculty.

References

1. Commission on Professionals in Science and Technology (CPST). *Professional Women & Minorities: A Total Human Resource Data Compendium*; CPST: Washington, D.C., 1997 and 2004.
2. Eccles, J.; Jacobs, J. E. *Signs: Journal of Women in Culture and Society.* **1986**, 11, 367-380.
3. Andrews, F. In *Scientists in Organizations*; Pelz, D.; Andrews, F., Eds.; Institute for Social Research: Ann Arbor, MI, 1976; pp 337-365.
4. Cole, J.; Cole, S. *Social Stratification in Science*; University of Chicago Press: Chicago, IL, 1973.
5. Glynn, M. A. *The Academy of Management Review.* **1996**, 21, 1081-1111.
6. Damanpour, F. *Academy of Management Journal.* **1991**, 34, 555-590.
7. Cole, J. *Fair Science: Women in the Scientific Community*; Free Press: New York, NY, 1979.
8. Fox, M. F. In *Handbook of Science and Technology Studies;* Jasanoff, S.; Markle, J.; Petersen, J.; Pinch, T., Eds.; Sage Publications: Newbury Park, CA, 1995; pp 205-223.
9. National Research Council. *Climbing the Ladder: An Update on the Status of Women Scientists and Engineers;* National Academy Press: Washington, D.C., 1983.
10. Hornig, L. S. In *Women: Their Underrepresentation and Career Differentials in Science and Engineering*; Dix, L. S., Ed.; National Research Council: Washington, DC, 1987; pp 103-122.

11. Bruer, J. *Science, Technology, and Human Values.* **1984**, *9*, 3-7.
12. Ahern, N.; Scott, E. *Career Outcomes in a Matched Sample of Men and Women Ph.D.s*; National Academy Press: Washington, D.C., 1981.
13. Long, J. S.; Allison, P.; McGinnis, R. *American Sociological Review.* **1993**, *58*, 703-722.
14. Astin, H.; Davis, D. In *Scholarly Writing and Publishing: Issues, Problems, and Solutions;* Fox, M. F., Ed.; Westview: Boulder, CO, 1985; pp 147-160.
15. Cole, J.; Zuckerman, H. *Scientific American.* **1987**, *255*, 119-125.
16. Fox, M. F. *Social Studies of Science.* February **2005**, *35*.
17. Helmreich, R., et al. *Journal of Personality and Social Psychology.* November **1980**, *39*, 896-908.
18. Reskin, B. *American Journal of Sociology.* **1978**, *83*, 1235-1243.
19. Long, J. S. *Social Forces.* **1990**, *68*, 1297-1315.
20. Fox, M.F.; Faver, C. *The Sociological Quarterly.* **1985**, *26*, 537-549.
21. Xie, Y.; Shauman, K. *Women in Science: Career Processes and Outcomes*; Cambridge, Mass.: Harvard University Press, 2003.
22. Grant, L.; Kennelly, I.; Ward, K. *Women's Studies Quarterly.* Spring/summer **2000**, *28*, 62-83.
23. Fox, M. F. In *The Outer Circle: Women in the Scientific Community;* Zuckerman, H.; Cole, J.; Bruer, J., Eds.; W. W. Norton: New York, NY, 1991; pp 188-204.
24. Fox, M. F. In *Research and Higher Education: The United Kingdom and the United States;* Whiston, T. G.; Geiger, R. L., Eds.; The Society for Research into Higher Education & Open University Press: Buckingham, England, 1992; pp 103-111.
25. Fox, M. F. *Signs: Journal of Women in Culture and Society.* Autumn **1998**, *24*, 201-223.
26. Fox, M. F. *Women's Studies Quarterly.* Spring/Summer **2000**, *28*, 47-61.
27. Fox, M. F. *Gender & Society.* **2001**, 15, 654-666.
28. Fox, M. F. In *Equal Rites, Unequal Outcomes: Women in American Research Universities*; Hornig, L., Ed.; Kluwer Academic/Plenum Publishers: New York, NY, 2003.
29. Fox, M. F.; Stephan, P. *Social Studies of Science.* February **2001**, *31*, 109-122.
30. Fox, M. F. *Georgia Tech ADVANCE Survey of Faculty Perceptions, Needs, and Experiences*; Georgia Institute of Technology: Altanta, GA, 2003.
31. Mackie, M. *Sociology and Social Research.* **1977**, 61, 277-293.
32. Chubin, D. *American Sociologist.* May **1974**, *9*, 83-92.
33. Long, J. S. *Social Forces*. **1992**, *71*, 159-178.
34. Cameron, S. Ph.D. thesis, The University of Michigan, Ann Arbor, MI, 1978.
35. Fox, M. F. *Social Studies of Science.* **1983**, *13*, 285-305.
36. Merton, R. K. In *The Sociology of Science*; Storer, N. W., Ed.; The University of Chicago Press: Chicago, IL, 1973.

37. Mullins, N. C. *Science: Some Sociological Perspectives*; Bobbs-Merrill: Indianapolis, IN, 1973.
38. Creamer, E. *Assessing Faculty Publication Productivity: Issues of Equity*; ASHE-ERIC Higher Education Report Volume 26, No. 2; The George Washington University: Washington, D.C., 1978.
39. Long, J. S.; Fox, M. F. *Annual Review of Sociology.* **1995**, *21*, 45-71.
40. Long, J. S. *From Scarcity to Visibility: Gender Differences in the Careers of Doctoral Scientists and Engineers;* National Academy Press: Washington, D.C., 2001.
41. Sonnert, G.; Holton, G. *Who Succeeds in Science? The Gender Dimension;* Rutgers University Press: New Brunswick, NJ, 1995.
42. Fox, M. F. In *Handbook of the Sociology of Gender;* Chafetz, J. S., Ed.; Kluwer Academic/Plenum Publishers: New York, NY, 1999, pp 441-457.
43. Rossiter, M. *Women Scientists in America: Struggles and Strategies to 1940*; Johns Hopkins University Press: Baltimore, MD, 1982.
44. Rossiter, M. *Women Scientists in America: Before Affirmative Action, 1940-1972*; Johns Hopkins University Press: Baltimore, MD, 1995.
45. Fox, M. F. In *The Equity Equation: Fostering the Advancement of Women in the Sciences, Mathematics, and Engineering;* Davis, C. S.; Ginorio, A.; Hollenshead, C.; Lazarus, B.; Rayman, P., Eds.; Jossey-Bass: San Francisco, CA, 1996, pp 265-289.
46. Fox, M. F. *The Annals of the New York Academy of Sciences.* April **1999**, *869*, 89-93.
47. Feldt, B. *An Analysis of Productivity of Nontenured Faculty Women in the Medical and a Related School: Some Preliminary Findings;* Office of Affirmative Action: Ann Arbor, MI, 1985.
48. Feldt, B. *The Faculty Cohort Study: School of Medicine;* Office of Affirmative Action: Ann Arbor, MI, 1986.

Notes

---

[i] Scientific fields comprise the eight classifications of the National Science Foundation and National Research Council: physical, mathematical, computer, environmental, life, and engineering, as well as the psychological and social sciences.

[ii] Men with doctoral degrees in science (44%) are somewhat less likely than women (52%) to be in educational compared to other sectors. The gender difference owes principally to the higher concentrations of men in engineering fields which are, in turn, more likely than other science fields to be practiced in nonacademic, industrial settings.

[iii] The classifications here include earth, atmospheric, and ocean sciences, and mathematical/computer sciences, as well as physics and chemistry.

[iv] Microbiology students were also surveyed, but not included in the summary here. This is because for microbiology, the field in which students identify their degree (in National Research Council's Survey of Doctoral Recipients) corresponds more loosely with department, and cannot be sampled with the same design of departments that have been low, high, or improved in doctoral degrees awarded to women. A degree in microbiology may be from variable departments, such as molecular genetics, neurobiology, or other units.

Chapter 4

Gender Disparity in the Training and Mentoring of Chemists

Susan A. Nolan

Department of Psychology and Center for Women's Studies, Seton Hall University, 400 South Orange Avenue, South Orange, NJ 07079 (email: nolansus@shu.edu)

This chapter will outline the importance of mentoring for chemists and others in STEM fields, and particularly for female chemists. It also will document the ways in which female chemists often fail to receive the mentoring necessary for educational and professional development. This chapter will outline the importance of mentoring female chemists at all levels of their training and careers, discuss research highlighting the differential mentoring experiences of women and men that often are detrimental to women, suggest individual and institutional practices that might effect change, and highlight possible avenues for future research.

Numerous studies have documented the importance of mentoring to educational and professional success (e.g., *1, 2, 3*). Certainly, mentoring seems essential to success in STEM fields, including chemistry, particularly for women (*1*). Yet, despite the documented importance of mentoring, women appear to receive less mentoring, and to perceive the quantity and quality of their mentoring differently than do men (*4*). As will be outlined later in this chapter, gender differences in mentoring appear to play a major role in the progress toward gender equity in chemistry and other STEM fields.

As has been documented earlier in this book, there has been considerable progress toward gender equity in the labor force over the last several decades, yet women continue to be underrepresented at the highest levels across a range of fields (*5*). Among the fields that exhibit such a gender disparity are science, technology, engineering, and mathematics (STEM), in which the pattern of fewer women at each level in the hierarchy often is referred to as a "leaky pipeline" (e.g. 6). This phenomenon is evident in both education (*6, 7*) and careers (*8*). Within academia, the gender difference is especially pronounced in the most prestigious positions, tenure-track professorships at the top ten-ranked institutions, as identified by the National Research Council (NRC); although there have been slow, but gradual, increases in the percentages of women, the overall proportion of female faculty members in all types of positions remains quite low (*9, 10, 11*).

The field of chemistry exemplifies the problems faced by women in STEM fields. Despite a fairly large pool of women with doctorates from elite schools (*4*), women receive few of the most prestigious academic positions (*12, 13, 14*). In other words, despite the anecdotal complaint that there are not enough qualified women with doctorates who are available, the field of chemistry is hiring below the available pool of women. This is particularly problematic because there is a larger demand for STEM expertise than there is a supply (*15*), an indication that recruitment efforts will have to target women and minorities to meet our nation's need for STEM talent.

One of the best targets for increasing our nations STEM talent is the higher educational system. What aspects of education and training lead to lower hiring rates of women versus men? In terms of scientific training, it might seem as if women and men are having similar experiences; yet, there may be gender differences in the perceptions of these experiences. Rosser and Zieseniss (*16*), for example, report that some laboratory environments are viewed as intimidating by female engineers, perhaps because laboratory work might be viewed as "masculine" by both men and women. Such a view might lead women to develop anticipatory anxiety with respect to laboratory experiences, a view that might extrapolate to scientific careers more generally (*17, 18*). It is possible that other fields, including chemistry, would yield similar findings if studied.

Others have examined the harassment of women, noting that if students and faculty members fail to notice mistreatment of women, they might minimize the negative effects of such mistreatment (*19*). The ACS Early Careers of Chemists survey (*20*) reported that 3.5% of men and 45.9% of women responded that they had experienced discrimination based on gender, an indication that women do perceive mistreatment that is likely to go unnoticed by others, particularly by men. These findings highlight the importance of understanding women's and men's perceptions of their training experiences, particularly because perceptions might differentiate between those who are more and those who are less successful with respect to training, identity development as scientists, and employment.

Definition of Mentoring

First, it is important to define mentoring. Mentoring encompasses a variety of roles, and is more than merely teaching a student the skills necessary in a given field or micromanaging a student's research. Noe (*21*), for example, described the psychosocial, and not just the educational and career-related, benefits of mentoring. Schlosser and Gelso (*22*) studied the concept of a working reliance within advising relationships and found three factors: rapport (aspects related to the interpersonal relationship), apprenticeship (aspects related to professional development), and identification-individuation (aspects related to the process whereby an advisee determines how much like the advisor he/she wants to be). Moreover, he National Academy of Sciences, National Academy of Engineering, and Institute of Medicine (*1*) described the role of a mentor of science and engineering students as "adviser, teacher, role model, friend." In fact, this four-word description forms part of the title of their book. The authors are clear in their distinction between faculty advisor and mentor, noting that not all faculty advisors serve a mentoring role. In a general way, they see a mentor as someone who works with a student over time to help her or him become successful in her or his field. More specifically, they cite assistance with facets of education, discipline-specific socialization and networking, and employment as within a mentor's purview. Perhaps most importantly, they describe mentoring relationships as developing over time, evolving with the needs of students, and perhaps continuing well into students' eventual careers. When mentoring is referred to in this chapter, it is in reference to a relationship as described in the above research, rather than a relationship built solely on academic advising.

Mentoring: Same-Gender Pairings

Good mentoring is integral to good training (e.g., *23*), and may both occur differently for and be perceived differently by women and men. In all STEM fields, including chemistry, one's advisor is at the center of one's training experience. For example, in an interview study, both men and women reported the long-term benefits of positive mentoring experiences throughout training (*23*). Empirical examinations within a variety of fields (e.g., *24, 25*) have concluded that faculty mentors play a powerful role in undergraduate student development; studies cite outcomes that include increased student retention as well as improved academic performance. Although there are fewer studies of mentoring at the graduate level, the extant research also suggests the likelihood that faculty-student advising and mentoring play an essential role throughout one's academic career.

There are several studies within the mentoring literature, across a range of disciplines and levels of training that have examined specifically the role of female mentors for women. Jacobi (*24*), for example, noted that, "while none of the literature reviewed for this article flatly declares cross-sex or cross-race pairs to be completely unworkable, the problems of establishing and maintaining such relationships are described on a continuum ranging from mild to severe" (p. 511). Although some argue that gender is not relevant, most researchers in this area concur with Jacobi (e.g., *26 27*). Although the data are inconsistent with respect to the relative efficacy of same-gender vs. opposite-gender mentoring pairs, evidence does suggest both that students prefer same-gender mentors and faculty members are more likely to initiate such relationships with same-gender students (*27*). In fact, in a study of female and male scientists, approximately 23% of women had received mentoring from another woman, whereas about 2% of men had been mentored by a woman (*23*).

Erkut and Mokros (*28*) observed an interesting gender difference with respect to students' preferences in mentors. They reported that female students seek mentors based on availability, rather than specifically seeking female mentors. In other words, female mentors will be chosen by women in the percentages in which they are available. Conversely, these authors reported that men showed strong preferences for male mentors, even when female mentors were available. Gumbiner (*29*) reported that students perceived professors in similar ways. For example, students of both genders perceived male professors to be authoritative, viewed female professors as emotional, and preferred male professors in the classroom; despite these similarities in perceptions, both female and male students preferred same-gender mentors. Despite some small differences in findings from mentoring studies, there appears to be a general trend for women to prefer female mentors.

Does this preference for female mentors lead to particular benefits for women? Another line of mentoring research examined the possible benefits of same-gender mentors for female students. Overall, findings in this area suggest that such pairings do seem to be beneficial for women, or at least that women *perceive* that they are more beneficial for them. For example, Gilbert and Rossman (*30*) describe three benefits of mentoring derived by female students:

a) exposure to role models; b) acceptance and empowerment; and c) sponsorship by a mentor (i.e., introduction into the social systems of the academy). Female students themselves appear to be aware of these possible benefits. First, female students described role-model relationships, typically with same-gender mentors, as more important to their career development than did male students (*31*). Further, female students with same-gender mentors described themselves as more career-oriented, confident, and competent than did female students with opposite-gender mentors (*32*). In fact, female students' with female mentors described themselves in the same ways in which male students with male mentors described themselves. Moreover, female students with female mentors reported higher satisfaction as a student than did any student, regardless of gender, with a male mentor.

On the other hand, a preference for female mentors might be a detriment to women. Preston (*23*) observes two reasons that women might be less likely than men to have mentors throughout their training. First, she cites the possible preference of mentors for male trainees because of the historically documented success of men over the years. However, she also cites the preference of women for female mentors, which, because female mentors are rare, might land women in the position of not having a mentor at all. The implication is that women, although they might prefer a female mentor and even might receive better mentoring from a female mentor, must be actively open to possible mentors of either gender.

Differential Experiences of Mentoring

In addition to a likely preference for female mentors among women, women also appear to experience the mentoring relationship differently than do men, regardless of the mentor's gender. Specifically, research has suggested that mentoring has a different effect on women than on men. Preston (*23*) discusses the ironic finding that despite the fact that men receive more mentoring in their training than do women, it is less important to their career success than it is to the career success of women, particularly in the short-term. For example, Preston documented that only 13.5% of women in her study said that they had mentors as undergraduates, compared to 40.0% of men, and 20.5% said that they had mentors in graduate school, compared with 65.7% of men. Yet, the effect of these mentors was greater for women. Women's chance of finishing graduate school increased from 60% to 100% with a strong mentor, whereas a strong mentor had no effect on men (75% to 74%). Similarly, women's chance of successfully obtaining employment increases from 52% to 100% with a mentor, whereas men's chance increases only from 70% to 83% with a mentor. Preston's work also documented that among men and women who *never* had a strong mentoring relationship, women fared far more poorly than did men with respect to success and longevity in the field of science. The premise that mentoring is more valuable to women than to men is bolstered by theoretical work by Athey, Avery, and Zemsky (*33*) who discussed mentoring as a way in which one can increase one's "human capital." Athey et al. contended that women's decreased access to informal interactions, primarily because of their minority status within science, makes the formal mentoring situation that one experiences with an advisor essential to their acquisition of human capital.

In order to diagnose the mentoring-related problems that women might face throughout their training, it is imperative that one understands the ways in which women and men differentially experience their training. Three colleagues and I recently conducted a large-scale survey study of 455 men and women who graduated from NRC-ranked top ten chemistry doctoral programs between 1988 and 1992 (*4*). This study examined participants' perceptions of mentoring at the undergraduate, graduate, and post-doctoral levels and found a number of gender differences. The full study is reported elsewhere, but the highlights of the results are summarized here. It must be noted that many of the questions were about advisors, rather than mentors specifically; therefore, some responses might be about advisors who were not mentors in the full sense of the word.

At the undergraduate level, women were less likely to have attended institutions with a graduate program; this finding suggests that women had less access than did men to mentors who were privy to informal networks at top Ph.D. programs. Moreover, although men and women were similarly likely to have had research experiences as undergraduates, men were more likely than were women to state that an undergraduate professor had helped them find a research experience. Women, on the other hand, were more likely than were men to state that they found their research experiences through formal channels, such as postings within their academic programs.

When asked who had helped them to choose a graduate school, men were significantly more likely than were women to cite the help of an undergraduate professor (the most common response for both genders), whereas women were more likely than were men to respond either "myself" or "no one." Interestingly, women were more likely than were men to say that a woman helped them choose a graduate school, again highlighting the benefits of same-gender mentors for women. Once in graduate school, men gave higher scores than did women when asked to rate how well-prepared they were for graduate school as compared to other members of their research groups.

When queried about experiences once in graduate school, women were more dissatisfied with mentoring than were men. Although there were few reported gender differences with respect to informal networking, women perceived that they received less formal mentoring than did men. For example, although men and women used many of the same criteria to select a mentor, women were less satisfied with their choice of criteria than were men and were more likely to report that they actually had switched to new advisors during their graduate training. In addition, men, compared to women, reported receiving more help in selecting an advisor from research and work supervisors, as well as their institutions' administrators, faculty members, and post-doctoral fellows.

Women's dissatisfaction with their graduate advisors also is evident in ratings of their interactions with their advisors. Men gave higher ratings than did women to their overall interactions with their dissertation advisors, the interest their dissertation advisors showed in them, and the help they received from advisors across a wide range of areas (e.g., formulation of research goals, access to appropriate equipment). In particular, three areas in which men reported receiving more help than did women – developing career goals, identifying personal motivators, and finding a job – involve more than learning specific skills, and thus, likely would require the development and maintenance of a personal relationship with one's advisor beyond a basic "academic" or curricular one.

Many of the gender patterns observed at the graduate level also emerged at the post-doctoral level. As before, there were few gender differences with respect to perceptions of informal networking, whereas there were several differences with respect to formal mentoring. First, when choosing a post-doctoral advisor, significantly more men than women reported receiving advice from their dissertation advisors. Moreover, among those who received such help, men gave higher ratings than did women to the quality of this assistance. This discrepancy is a particular detriment for women because of the repercussions of this choice for one's career. The Committee on Science, Engineering, and Public Policy (*34*) directly spoke to the importance of this choice; "The decision about whether to undertake a postdoctoral appointment is seldom easy and should involve consultation with one's advisor and as many mentors or other experienced contacts as possible" (p. 21). Women clearly are at a disadvantage compared to men if they are less likely to seek or receive such advice.

Within the post-doctoral relationship, men continued to report preferential treatment more than did women. First, men viewed their overall advisor/advisee interactions in a better light than did women. For example, in response to specific survey items, men reported more advisor interest than did women in their findings, research ideas, and publications, and thought that they had more publication opportunities than did women.

Finally, we examined mentoring within respondents' initial employment. There were again few gender differences reported with respect to informal mentoring; however, as at the post-doctoral level, more men than women reported that their dissertation advisors helped them find their first position. Once in their positions, men gave higher ratings to the support that they received from department chairs and supervisors than did women. Thus, there is some indication that women continue to be at a disadvantage with respect to formal mentoring within their careers more so than do men. Because the pattern of qualitatively and quantitatively poorer formal mentoring for women than for men holds throughout training and initial employment, our data point to the ongoing nature of the disadvantage that women differentially experience, which in turn may translate into additional disadvantages at each ensuing level of training and career.

The Lack of a Critical Mass of Female Mentors

The research cited previously has documented the importance of mentors, and particularly of female mentors, for women. In addition, research suggests that women and men experience mentoring differently. A number of studies have explored possible reasons for gender differences in perceptions of mentoring, focusing on gender differences in behaviors of the mentors themselves. For example, Fox (*35*) attributed differences in perceptions to actual differences in male and female mentoring styles. Fox suggested that female faculty members approach the task of mentoring more broadly than do male faculty members in terms of both content (i.e., what they discuss in the mentoring relationship) and in approach (i.e., how they construct the mentoring relationship). As an example, female faculty members are more likely than are male faculty members to emphasize participation in research group meetings, frequent interaction with faculty, and acquisition of a wide range of skills rather than honing in on very specific specialty skills. As another specific example, female mentors also set higher standards for their female students in seminar presentations than do male mentors. It is possible that different mentoring styles might lead to the acquisition and development of different sets of skills with different implications for future training, as well as for career decisions and opportunities (*4*).

If women indeed have different, and perhaps more beneficial, styles of mentoring, it would follow that the presence of female mentors would improve the education and career prospects of women. It is not surprising, then, that within STEM fields, perhaps the most commonly cited reason for the mentoring deficit experienced by women is the absence of a "critical mass" of senior women (e.g., *11*). Such a cadre of women could mentor graduate students, post-doctoral fellows, and female junior faculty, and alleviate women's current diminished access to the support systems and resources that are so necessary to succeed in academia (*36, 37*). For example, mentors are essential to the establishment of network contacts with recognized authorities in one's field. If women lack good mentoring, they do not have critical allies advocating their promotion, tenure, and nomination for awards (*38*). Recent data document the absence of a critical mass of women in the field of chemistry specifically.

Currently there are, on average, four female faculty members out of an average of 33 faculty members per department among the top 50 NRC-ranked chemistry departments. In fact, twelve of these departments have two or fewer female faculty members (*14*). Clearly, chemistry suffers from a shortage of senior female faculty members.

Detriments of Differential Mentoring and of a Lack of Female Mentors

A recent study (*23*) compared pairs of individuals who had been trained as scientists; each woman or man who had left the field of science was paired with a same-gender individual who had chosen to stay in the field of science. The results clearly demonstrated the greater importance of mentoring for women than for men. For each pair, the author attempted to identify a factor that differentiated the one who left from the one who stayed. For women, the second most frequently identified differentiating factor – in seven out of twenty-two pairs – was "lack of mentor or guidance" (*23*, p. 32). In six of the seven pairs, the woman who left science did so because of the lack of a mentor specifically at the undergraduate or graduate level. The lack of mentoring was not a differentiating factor in any of the 19 pairs of men whom Preston studied, clear evidence that lack of mentoring can precipitate directly women's departure from the field of science, but does not appear to do so for men. Interestingly, the author noted that sex discrimination and gender-based double standards were only indirect factors in women's decisions to leave insofar as they led to fewer mentors for women than for men. Preston points out that one could reframe these findings, stating that women who stayed were the recipients of strong mentoring; in fact, she notes that every woman in her study who could describe a time in her career during which she had a strong mentor could also describe at least one satisfying scientific experience. Preston also discusses the concept of an "anti-mentor," an individual who actually thwarts an individual's attempts to succeed, and reported that anti-mentors were more common among the women in the cohort that she studied than among the men (*23*, p. 102). In response to the possible contention that failure in mentoring relationships was the fault of the trainee as opposed to the mentor, Preston stated that her study uncovered numerous examples of women who followed a failure experience with a mentor with a success experience with a different mentor.

As already discussed, mentoring is perhaps so important for women because of their overall lower access to the information necessary to succeed in science; without a strong mentor, students are deprived of numerous benefits. For example, Preston (*23*) discusses the role of a mentor in helping a student to learn scientific skills, network with other professionals, and obtain desirable employment. Gibbons (*39*) more specifically cites the importance of mentors' help to students by connecting them with professionals in the field who organize conferences and can ease students' road to inclusion in professional meetings, as well as with journal editors who can ameliorate the publication process. Rose reported that a lack of mentoring is likely the cause of the documented decreased tendency of junior female faculty to maintain active ties with their doctoral institutions (*40*), a factor in women's smaller professional networks; she attributes this tendency directly to the connection or lack thereof with one's doctoral dissertation advisor.

Gender also has been shown to be a factor in collaboration between a student and an advisor. Long (*41*) documented that female students who have young children have lower rates of research collaboration with their advisors than do female students who do not have young children. Long notes that there is no such difference among men who do and do not have young children. Long and McGinnis (*42*) also cite a positive correlation between an advisor's prestige and a student's success in graduate school, post-doc, and career; this correlation is strongest when students and advisors collaborate on research. Because women are a) more likely than are men to have female advisors, who have, in general, received less recognition, and b) are more likely than are men to have advisors with lower academic rank, the correlation between advisor prestige and trainee success does not work to the benefit of women (*41*).

Research discussed thus far suggests both that women tend to receive lower quality mentoring than do men and that women may be less connected to the academic networking opportunities to which one is typically introduced through one's advisor. It is possible that the exclusion of women from such settings might lead to perceptions of women as less qualified researchers and leaders than men. Indeed, Eagly, Karau, and Makhijani (*43*) reported that in both professional and informal settings that are dominated by men, women are viewed more negatively and as having less effective leadership styles than are men (c.f., *44*). Cultural gender schemas, or cognitive expectations, include beliefs about men's and women's competence and abilities, as well as assumptions about the careers deemed appropriate for women and men (*5*). It is possible that such gender schemas directly lead to differences in what is expected for men and women in careers, as well as perceptions about how men and women perform in career roles. Many people perceive that traditional female roles are in conflict with the reality of the many women who work full-time. If gender-based stereotypes shape beliefs about self and performance, they also might affect individuals' levels of self-efficacy and goal-oriented motivation. Thus, research on gender schemas further enhances the importance of perceptions and realities of the mentoring within training environments. The views one holds about one's environment can directly shape one's developing identity as a scientist, one's motivation to achieve, and ultimately, one's success within a scientific career.

It is important to note that a number of the statistically significant gender differences observed in the studies within the mentoring literature, including our own study, are small effects according to statistical conventions for effect sizes. Thus, although the differences *are* statistically significant and, therefore, likely exist, many of them are small in size. One may not be able to perceive these differences readily in the training and employment contexts in which chemists work, perhaps a reason that so many perceive gender equity in training and career experiences, including mentoring. It is important to note, however, that the fact that many of women's disadvantages are "small" and often cannot be observed easily likely accounts for the fact that many deny such gender differences. Moreover, this likely also accounts for the fact that many view any enforced institutional change as unnecessary and even detrimental to the scientific environment. Small differences, however, do have effects – often large ones – on both training and long-term career trajectories, particularly because the small differences are numerous and pervasive. In her writings on the reasons that women continue to lag behind men across most professional fields, Valian (*5*) champions the sociological theory referred to as the "accumulation of advantage and disadvantage" (e.g., *45*). Valian explains that this theory "suggests that, like interest on capital, advantages accrue, and that, like interest on debt, disadvantages also accumulate" (*5*, p. 3). Valian cites myriad empirical findings highlighting the career damage that a series of small slights to women can inflict. She convincingly supports her case that even the smallest differences in favor of men must be eliminated; as she succinctly states, "mountains *are* molehills, piled one on top of the other" (italics in the original; *5*, pp. 4-5). Within the field of chemistry, it is likely that small gender differences in mentoring at the training and employment levels translate into differential success (e.g., lower doctoral achievement rates among women, fewer women on tenure-track faculty at research institutions). For example, we examined the pool from which the participants in our own study were taken – men and women who graduated from top ten doctoral institutions between 1988 and 1992. Within this population, men were 1.8 times more successful than were the women in obtaining a tenure-track faculty position in 2001 at a top 50 school (*4*). As my colleagues and I have observed, this disparity, although not always readily observable, is startling to those working or training in this context when they confront the data demonstrating this gender gap.

Implications for Change

The existing literature on mentoring offers several prescriptions for improvement on both the individual and institutional levels. First, on the individual level, undergraduate, graduate, and post-doctoral female trainees, as well as their peers and their potential mentors, must become aware (or be made aware) of the ways in which female students and post-doctoral fellows are likely to be excluded from formal and informal interactions. Enhanced awareness might encourage women to be more active in the mentoring process (e.g., actively garnering information about possible advisors, seeking or forming support groups). For female and male faculty members, awareness that their female graduate students might perceive that they are excluded might lead to more active inclusion of women in post-doctoral and career guidance, the establishment of networking connections, and active encouragement of women to become involved in formal and informal networks.

Of equal, if not greater, importance is the task of helping female trainees and their mentors to realize that there are likely incongruities between female trainees' perceptions of training and mentoring and those of their mentors, who are predominantly male. This discrepancy has direct consequences because mentoring relationships not only provide the context in which one learns about one's fields from experts, but also serves as a paradigm for interactions within the scientific community. A strong mentoring relationship can help trainees to learn effectively about, put into practice, and receive appropriate feedback on the skills necessary for development as a scientist. In addition, it is important to help female trainees and their mentors to realize that they might have different expectations about their advisors' roles than do their male counterparts; such cognitions can and *do* shape their attributions regarding subsequent successes and/or failures. Specifically, the failure to realize that one's mentoring is qualitatively different from that of one's peers might lead one to blame oneself for failure, when really it is more a failure of the mentoring practice.

At the institutional level, it is important to increase departmental and university awareness of gender differences in perceptions of mentoring so that institutions might implement more formal mentoring and advising programs at all levels. For instance, formal support groups or mentoring programs for women, or for both women and men, might take the form of senior-junior mentoring partnerships, tenure and promotion workshops in academia, research and writing forums, financial and non-monetary incentives for collaborative initiatives and grant proposals, and introduction to broader networks within the field. Preston (*23*) specifically advocates multi-level mentoring for women in which post-doctoral students mentor graduate students who in turn mentor undergraduates. She credits the ensuing network of women as a way to share access to the larger, more established professional networks, and notes that multi-level mentoring also might include men. She encourages the inclusion of social occasions in multi-level mentoring to strengthen the ties among the women in this network. Although Preston's model could occur naturally in multi-level research groups, without an explicit format, the communication that fosters a mentoring environment may not occur. In line with a call for structure, Athey,

Avery, and Zemsky (*33*) outline the need for mandated formal mentoring programs because informal mentoring tends to focus on the short-term, a strategy that leads members of the majority to receive the most guidance. Only with mandated mentoring, or in a situation in which good mentoring is explicitly rewarded, can institutions achieve the long-term benefits that diversity offers.

Institutions also might introduce more active and structured "marketing" of their strong female candidates so as to increase their viability in the upper echelons of the post-doctoral, academic, and industrial job markets. The more that institutions succeed in the recruitment and marketing of women, the more women will be available to serve as role models and mentors for succeeding generations of scientists – both female and male. Regardless of the changes that institutions or departments aim to implement, many researchers in this area insist that the upper levels of the hierarchy of the particular department or institution must explicitly support the change if it is to succeed (e.g., *23*).

Directions for Future Research

At the undergraduate level, institutions might initiate research to explore whether women benefit from attending four-year colleges, as opposed to universities. Because female faculty members are more common at such schools (*44*) than at other types of institutions, it would follow that students would find female mentors and role models more readily at such institutions. It remains for further research to assess whether this results in enhanced success for women in graduate school.

Future research also should examine the outcome of the implementation of developmental and institutional initiatives like those outlined above. The field would benefit from comparisons of departments and institutions that have and have not developed informal or formal programs, as well as from experimental studies in which female graduate students are assigned randomly to different types of support programs. In addition, my colleagues and I are in the process of expanding our own survey research to include graduates of top doctoral programs in other STEM fields, including chemical engineering, electrical engineering, mathematics, and physics. A nuanced understanding of how gendered patterns of perception differ across STEM fields might illuminate techniques that forward the goal of equity for women and men across all types of scientific and technical domains and those that are uniquely confined to specific disciplines. For example, it is important to develop an understanding of why *only* chemistry, among these fields, is hiring below the available pool of qualified women, despite indications that these women are applying for positions in high numbers (*45*).

Summary

In summary, the existing literature on mentoring in general, as well as in the field of chemistry more specifically, suggests that women are not receiving the same quantity and quality of mentoring as are men. Previous studies by others have shown that, for women in particular, retention and success in the field correlates with the presence of good mentors. Formal and informal intervention by individuals and institutions clearly is necessary to redress this gap. Although there is an existing belief among many in STEM fields that women have achieved parity with men, data suggest that women and men are differentially successful in training and careers. It is clear that interventions aimed at mentoring and other training-related factors are necessary to further reduce gender disparities in success both for female and male chemists, and for women and men in STEM fields more generally.

References

1. National Academy of Sciences, National Academy of Engineering, and Institute of Medicine. *Adviser, teacher, role model, and friend: On being a mentor to students in science and engineering;* National Academy Press: Washington, DC, 1997.
2. Hughes, M. S. *Empowering women: Leadership development strategies on campus. New directions for student services: No. 44;* Developing leadership potential for minority women. In M. A. D. Sagaria, Ed.; Jossey-Bass, San Francisco, CA, 1988; pp 63-75.
3. Ugbah, S.; Williams, S. A. *Blacks in higher education: Overcoming the odds;* The mentor-protégé relationship: Its impact on blacks in predominantly white institutions; Elam, Ed.; University Press of America: Lanham, MD, 1989; pp 29-42.
4. Kuck, V. J.; Marzabadi, C. H.; Nolan, S. A.; Buckner, J. P. *J. Chem. Ed.* **2004,** *81, 356-363.*
5. Valian, V. *Why So Slow? The Advancement of Women;* The MIT Press: Cambridge, MA, 2000.
6. Kuck, V. J. *Chemical & Engineering News.* **2001**, *79, 71-73.*
7. National Science Foundation, Division of Science Resources Statistics. *Science and engineering doctorate awards: 2002;* National Science Foundation: Arlington, VA, 2002.
8. National Science Foundation, Division of Science Resources Statistics. *Characteristics of doctoral scientists and engineers in the United States: 2001;* National Science Foundation: Arlington, VA, 2001.

9. Long, J. S. *From scarcity to visibility: Gender differences in the careers of doctoral scientists and engineers*; National Academy Press: Washington, DC, 2001.
10. Long, J. S.; Fox, M. F. *Annual Review of Sociology.* **1995**, *21, 45-71.*
11. Sharpe, N. R.; Sonnert, G. *Journal of Women and Minorities in Science and Engineering.* **1999**, *5, 207-217.*
12. Brennan, M. B. *Chemical & Engineering News.* **1996**, *74, 6-16.*
13. Byrum, A. *Chemical & Engineering News.* **2001**, *79, 98-103.*
14. Marasco, C. A. *Chemical &Engineering News.* **2003**, *81, 58-59.*
15. National Science and Technology Council. *Ensuring a strong U.S. scientific, technical, and engineering workforce in the 21st century;* National Science and Technology Council: Washington, DC, 2001.
16. Rosser, S. V.; Zieseniss, M. *Journal of Women and Minorities in Science and Engineering.* **2000**, *6, 95-114.*
17. Conefrey, T. *Journal of Women and Minorities in Science and Engineering.* **2000**, *6, 251- 264.*
18. Traweek, S. *Beamtimes and lifetimes: The world of high energy physics*; Harvard University Press: Cambridge, MA, 1988.
19. Fox, M. F. *Women's Studies Quarterly: Special Issue, Building Inclusive Science: Connecting Women's Studies and Women in Science and Engineering.* **2000**, *28, 47-61.*
20. American Chemical Society. *The early careers of chemists: A report on the American Chemical Society's study of members under age 40*; American Chemical Society: Washington, DC, 2002.
21. Noe, R. A. *Personnel Psychology.* **1988**, *41*, *457-479.*
22. Schlosser, L. Z.; Gelse, C. J. *Journal of Counseling Psychology.* **2001**, *48, 157-167.*
23. Preston, A. E. *Leaving science;* Russell Sage Foundation: New York, 2004.
24. Jacobi, M. *Review of Educational Research.* **1991**,*61*, *505-532.*
25. Thile, E. L.; Matt, G. E. *Journal of Multicultural Counseling and Development.* **1995**, *23*, *116-126.*
26. Erkut, S.; Mokros, J. R. *American Educational Research Journal.* **1984**, *21, 399-417.*
27. Gumbiner, J. *Psychological Reports.* **1998**, *82, 94.*
28. Gilbert, L. A.; Rossman, K. M. *Professional Psychology: Research and Practice.* **1992**, *23, 233-238.*
29. Gilbert, L. A. *Sex Roles.* **1985**, *12, 111-123.*
30. Gilbert, L. A.; Gallessich, J. M.; Evans, S. L. *Sex Roles.* **1983**, *9, 597-607.*
31. Athey, S.; Avery, C.; Zemsky, P. *American Economic Review.* **2000**, *90, 765-786.*

32. Committee on Science, Engineering, and Public Policy of the National Academy of Sciences, the National Academy of Engineering, and the Institute of Medicine. *Enhancing the postdoctoral experience for scientists and engineers: A guide for postdoctoral scholars, advisers, institutions, funding organizations, and disciplinary societies;* National Academy Press: Washington, DC, 2001.
33. Fox, M. F. *Equal rites, unequal outcomes: Women in American research universities;* Gender, faculty, and doctoral education in science and engineering. In L. S. Hornig, Ed.; Kluwer Academic/Plenum Publishers: New York, 2003; 91-109.
34. Riordan, C. A.; Manning, L. M.; Daniel, A.M.; Murray, S. L.; Thompson, P. B.; Cummins, E. *Journal of Women and Minorities in Science and Engineering.* **1999**, 5, 29-52.
35. Stansbury, K. *The relationship of the supportiveness of the academic environment to the self-confidence and assertiveness in academic work for men and women graduate students in science and engineering.* Paper presented at the annual meeting of the American Educational Research Association: San Francisco, CA, 1986.
36. Farley, J. *Women professors in the USA: Where are they?;* Storming the tower: Women in the academic world; In S. Lie and V. O'Leary, Eds.; Nichols/Kogan Page: New York, 1990; pp 194-207.
37. Gibbons, A. *Science.* **1992**, *255, 1368-1369.*
38. Rose, S. M. *Psychology of Women Quarterly.* **1985**, *9, 533-547.*
39. Long, J. S. *Sociological Forces.* **1990**, *68, 1297-1315.*
40. Long, J. S.; McGinnis, R. *Scientometrics.* **1985**, *7, 255-280.*
41. Eagly, A H.; Karau, S. J.; Makhijani, M. G. *Psychological Bulletin.* **1995**, *111*, 3-22.
42. Kite, M. E.; Russo, N. F.; Brehm, S. S.; Fouad, N. A.; Hall, C. I.; Hyde, J. S.; Keita, G. P. *American Psychologist.* **2001**, *56, 1080-1098.*
43. Cole, J.; Singer, B. *The outer circle: Women in the scientific community;* A theory of limited differences: Explaining the productivity puzzle in science; In H. Zuckerman, J.R. Cole & J.T. Bruer, Eds., W.W. Norton: New York, 1991; pp 277-310.
44. American Chemical Society. *ChemCensus 2000: Analysis of the American Chemical Society's comprehensive 2000 survey of the salaries and employment status of its domestic members;* American Chemical Society: Washington, DC, 2000.
45. Marzabadi, C. H.; Kuck, V. J.; Nolan, S. A.; Buckner, J. A.. Career outcomes of doctoral graduates from top-ranked chemistry departments: Results from a career continuity study. Unpublished manuscript, **2004**.

Chapter 5

Institutional Barriers for Women Scientists and Engineers

What Survey Data of NSF Professional Opportunities for Women in Research and Education and Clare Boothe Luce Professorship Awardees Reveal

Sue V. Rosser

Ivan Allen College, Georgia Institute of Technology, Atlanta, GA 30332–0525

In an effort to better understand the barriers and discouragements encountered by female faculty in the sciences and engineering, this chapter analyzes research comparing the experiences of Professional Opportunities for Women in Research and Education (POWRE) awardees and Clare Boothe Luce (CBL) Professorship recipients. Responses of awardees underline the need for institutional, systemic approaches to overcome the obstacles.

Introduction

"I apologize for not writing sooner and responding to your questions. In fact, I'm not sure that I can respond to your first two questions in an objective way. I am experiencing a painful situation in my professional life and find I'm unable to write about it. Perhaps this situation is related to challenges

> facing women scientists in general or perhaps it is my individual experience. Nothing like this has happened to me before. I would be willing to speak with you over the phone and would appreciate the opportunity to do so. You can decide whether the information I provide is relevant and a reflection of the situation for women scientists in general or the institution where I'm located. Thank you." –Sharon, research scientist from a prestigious Research I University on the West Coast.

After receiving her Ph.D. and completing two post-docs in immunology, Sharon had struggled with keeping her career as a bench scientist. After the birth of her two children, she had worked part-time and then full time in a group with about 8 other individuals. The head of the group told her that it was time for her to become independent. Based upon his suggestion, she wrote and obtained the NSF grant. Although the project funded was related to the work she had done in the group, suddenly she found herself marginalized from the group, who no longer wanted to discuss results with her, and moved to a different location, with all major equipment, including her printer, removed. Although her grant paid her salary and some expenses, she needed the group for equipment and more importantly for collegial interaction and support. She has been made to feel unwanted with no control over her space and equipment. Although individuals in other institutions seem interested in her work, she cannot move because of family constraints. She is wondering whether she can still pursue her career as a scientist, despite the fact that she has a grant to support the work. As the quotation above suggests, she also wonders whether this is something only she faces at her individual institution or whether these circumstances commonly plague women scientists throughout academia.

In response to that question, I was not surprised to hear that other senior women scientists are frustrated and thinking of dropping out of science or switching to something else.

> "Being at a small liberal arts college, we're not as isolated as some of my colleagues from graduate school who went to research institutions. I did manage to have a family, and I still enjoy teaching, although prepping my own labs, the large number of contact hours semester after semester, and all the committee work I get because I'm a woman, have left me pretty burnt out. Still, I notice that most of my male colleagues have managed to keep their research going, at least at some low level, but mine went by the way several years ago when my kids were little. I regret it though and wonder sometimes if there is any way I could get it back. I wish there

> were some way that either the college could help me or maybe there is some program sponsored by a foundation that might make this possible for me and the other women. If not, I'm not sure how long the others and I can hang in here." –Jane, a tenured full professor at a small, Northeastern prestigious liberal arts college.

These women ask: Is it my individual failing as a woman and a scientist that makes me question the possibility that I can have a successful, happy career in academia? Are the problems I'm having the result of barriers that most women scientists at my institution (and most institutions) face as we try to build reasonable personal and professional lives?

As a dean at a Research I institution and as a scholar who has worked for a quarter of a century on theoretical and applied problems of attracting and retaining women in science and engineering, I have heard the expression of these doubts and dilemmas in a variety of forms from diverse female scientists and engineers in all types of institutions. Virtually all of the women are united in their love for science and desire to sustain their interest in the physical, natural world that attracted them to the study of science initially. Most would like nothing better than to pursue that love through their research and teaching in academia. But as the women themselves know, and as the statistics about gender and science document, more women than men are lost from science at every level of the pipeline. The female scientists question whether their individual choices, decisions, and will power, or institutional obstacles and barriers, prevent them from fulfilling their research potential and career goals.

In March, 1999, the Massachusetts Institute of Technology released "A Study on the Status of Women Faculty in Science at MIT", creating a stir that spread far beyond the institutional boundaries of MIT. Five years earlier senior biology professor Nancy Hopkins, *(1)* initiated the collection of evidence documenting that the 15 tenured female faculty members in science had received lower salaries and fewer resources for research than their male colleagues. Dean Robert Birgeneau recognized that in addition to salary disparities, the data in the report revealed systemic, subtle biases in space, start-up packages, access to graduate students, and other resources that inhibited the careers of female scientists relative to their male counterparts.

In January, 2001, MIT President Charles Vest hosted a meeting of the presidents, chancellors, provosts and twenty-five female scientists from the most prestigious research universities (California Institute of Technology, MIT, University of Michigan, Princeton, Stanford, Yale, University of California at Berkeley, Harvard, and the University of Pennsylvania). At the press conference held at the end of the meeting they recognized that barriers still exist for women and that "this challenge will require significant review of, and potentially significant change in the procedure within each university, and within the scientific and engineering establishments as a whole" (*2*, p.1).

Methods

In an effort to better understand the barriers and discouragements encountered by female faculty in the sciences and engineering, this chapter analyzes research comparing the experiences of Professional Opportunities for Women in Research and Education (POWRE) awardees and Clare Boothe Luce (CBL) Professorship recipients. The NSF established POWRE with two primary objectives: 1) To provide opportunities for further career advancement, professional growth, and increased prominence of women in engineering and the disciplines of science supported by NSF; and 2) To encourage more women to pursue careers in science and engineering by providing greater visibility for female scientists and engineers in academic institutions and in industry (*3*). POWRE awardees are women who received peer-reviewed funding from a focused National Science Foundation program from FY 1997-2000. They are primarily untenured assistant professors in tenure-track positions at research universities (RU I and RU II) as described by the Carnegie Classification of Postsecondary Institutions (*4*). The POWRE awards were capped at $75,000, with a typical duration of 12 to 18 months. A series of papers I wrote (*5, 6, 7, 8*) documents the research on 389 of the 598 POWRE awardees during the duration of the four-year NSF POWRE program (for the details of the POWRE program solicitation, see reference *3*).

The CBL Professorships were created by Clare Boothe Luce's generous bequest to The Henry Luce Foundation upon her death in 1987. One hundred and thirty-three women have been supported since then. CBL Professors are primarily assistant professors in their first tenure-track position at liberal arts colleges. Each CBL Professorship provides for the assistant or associate professor's salary, benefits, and a highly flexible career development account (generally $15,000-$20,000 annually in recent years) that is administered by the recipient. Support typically lasts for five years (for the details of the Clare Boothe Luce Professorships see reference *9*).

Because of the emergence of anecdotal reports that some female scientists actively choose to avoid research universities (*10*) because of their hostile climate it seemed important to query the Clare Boothe Luce Professors. Research universities have research and doctoral education as a primary part of their mission and expect faculty to publish research in reputable journals and attract peer-reviewed, competitive research funding to receive promotion and tenure. In contrast to research universities, four year institutions of higher education "are highly heterogeneous," including very prestigious liberal arts colleges, comprehensive institutions, and faith-based institutions. "What they have in common is that research and doctoral education is less central to their mission than is the case for research universities" (*11*, p. 126). Data supporting these anecdotes of women's avoidance of research universities documented that women make up 40% of tenure-track science faculty in undergraduate institutions (*12*), compared to less than 20% (*13*, Table 5-15) when data from four year colleges were combined with those from universities.

In order to examine this trend and to understand some of the reasons behind the data and anecdotal reports, the e-mail questionnaire and interviews administered to POWRE awardees (*6, 7*) were extended to female scientists and engineers concentrated at small liberal arts colleges. Although the NSF POWRE awardees included individuals from all types of institutions and at varying ranks, the overwhelming majority held the rank of untenured assistant professor and came from large research institutions. As reported in previous publications on this research (*7*), 67 of the 96 POWRE awardees for FY 1997, 119 of the 173 awardees for FY 1998, 98 of the 159 awardees for FY 1999, and 105 of the 170 awardees for FY 2000 to whom the e-mail survey was sent responded. The non-response rate ranged between 23% and 37% over the four-year period; the sample responding to the e-mail questionnaire in all four years appeared to be representative of the population of awardees with regard to discipline, and the non-respondents did not appear to cluster in a particular discipline. The limited data available from the e-mail responses revealed no other respondent or non-respondent bias.

The Clare Boothe Luce Professorships offered the survey a group of female scientists and engineers concentrated at small liberal arts colleges and private institutions who, like the POWRE awardees, had received an externally validated prestigious award. In the annual report information she collects from the current CBL professors, Jane Daniels, Program Director of the Clare Boothe Luce Professorships, included the same e-mail questionnaire that Rosser had sent to the POWRE awardees. Daniels also sent out the questionnaire to the former CBL professors. Forty-one of the forty-six active CBL professors responded to the questionnaire; eight of the eighty-four former CBL professors responded.

The two primary questions analyzed in this paper were the same ones previously reported on for the almost 400 POWRE awardees (*6, 7*):

1. What are the most significant issues/challenges/opportunities facing women scientists today as they plan their careers?
2. How does the laboratory climate (or its equivalent in your subdiscipline) impact upon the careers of women scientists?

Results

Table I lists the 17 categories into which the responses to question one by the women scientists and engineers were divided. The categories emerged from the coding of the textual replies (see *8* for further methodological details). The categories and data were discussed at a national conference by 30 social scientists, scientists and engineers whose work focuses on women and science (*14*). The same codes and categories were applied to the responses from the NSF POWRE awardees, as well as the responses from the CBL professors. Although most respondents replied with more than one answer, in some years at least one awardee gave no answer to the question. While the survey data are categorical and therefore not appropriate for means testing, differences in responses across award years clearly emerge when response frequencies are examined.

As Table I documents, the CBL professors give very similar responses to those of the POWRE awardees to question 1 about the most significant issues, challenges, and opportunities facing women scientists and engineers as they plan their careers. Even more strongly than their POWRE awardee counterparts, the CBL professors found "balancing career with family responsibilities" (response 1) to be the most significant issue. The CBL professors also ranked "low numbers of women, isolation, and lack of camaraderie" (response 3) and the "two career" problem (response 5) as significant issues, as had the POWRE awardees.

Contrary to the implications of the Schneider study (10), the responses of the CBL professors and POWRE awardees were remarkably similar, despite the differences in institutional size and type, with a few exceptions. CBL Professors ranked "time management/balancing committee responsibilities with research and teaching" (response 2) much lower than did the POWRE awardees. In fact, only one of forty-one current CBL professors mentioned this issue in the e-mail questionnaire responses. The likely reason that response 2 receives lower ranking from CBL professors than from POWRE awardees became evident from comments both in the e-mail questionnaire responses and in the interviews. The Clare Booth Luce Professors had the advantage of the flexibility of the CBL money (as opposed to the restrictions of federal money awarded through NSF's POWRE) that can be used to buy out teaching while establishing research, as well as for laboratory renovations, childcare, travel, hiring students, and other needs. "Gaining credibility/respectability from peers" (response 4) constituted another difference in responses between CBL professors and POWRE awardees. Approximately 20% of all 4 years of POWRE awardees cited "gaining credibility/respectability from peers and administrators" as a problem, while only about 10% of CBL professors cited this. Several of the POWRE awardees mentioned that since POWRE was an initiative for women only, many of their colleagues viewed the grant as less prestigious than other NSF grants, despite its very competitive success rate.

Looking more closely at some of the example quotations from the POWRE respondents from all four years and from the CBL Professors reveals the qualitative context for the categories and provides greater insight into the problems at hand. The women express the specific barriers for their careers:

Category A. Pressures Women Face in Balancing Career and Family

> "At the risk of stereotyping, I think that women generally struggle more with the daily pull of raising a family or caring for elderly parents, and this obviously puts additional demands on their time. This is true for younger women, who may struggle over the timing of having and raising children, particularly in light of a ticking tenure clock, but also for more senior women, who may be called upon to help aging parents (their own or in-laws). Invariably they manage, but not without guilt." (2000 POWRE respondent 63)

> "Child care benefits—I've never heard of anything similar elsewhere, and it's really a great way to make it easier for women in academia to balance work and family (not that it's ever easy)." (CBL respondent 37)

> "Managing dual career families (particularly dual academic careers). Often women take the lesser position in such a situation. Ph.D. women are often married to Ph.D. men. Most Ph.D. men are not married to Ph.D. women." (2000 POWRE respondent 16)

Category B: Problems Because of Low Numbers and Stereotypes

"Although possibly less now than before, women scientists still comprise a small proportion of professors in tenure-track positions. Thus, there are few "models" to emulate and few to get advice/mentoring from. Although men could also mentor, there are unique experiences for women that perhaps can only be felt and shared by other women faculty, particularly in other Ph.D. granting institutions. Some examples of this: a different (i.e., more challenging) treatment by undergraduate and graduate students of women faculty than they

Table I. Significant Issues Facing Women Scientists

Question 1: What are the most significant issues/challenges/opportunities facing women scientists today as they plan their careers?

	Categories	*1997 POWRE % of responses*		*1998 POWRE % of responses*		*1999 POWRE % of responses*		*2000 POWRE % of responses*		*Current CBL Profs. % of responses*		*Past CBL Profs. % of responses*		*Total CBL Profs. % of responses*	
1.	Balancing work with family responsibilities (children, elderly relatives, etc.)	62.7	(42/67)	72.3	86/119)	77.6	(76/98)	71.4	(75/105)	73.2	(30/41)	87.5	(7/8)	75.5	(37/49)
2.	Time management/balancing committee responsibilities with research and teaching	22.4	(15/67)	10.1	(12/119)	13.3	(13/98)	13.3	(14/105)	0.1	(1/41)	38.0	(3/8)	8.2	(4/49)
3.	Low numbers of women, isolation and lack of camaraderie/mentoring	23.9	(16/67)	18.5	(22/119)	18.4	(18/98)	30.5	(33/105)	26.8	(11/41)	---	---	22.4	(11/49)
4.	Gaining credibility/respectability from peers and administrators	22.4	(15/67)	17.6	(21/119)	19.4	(19/98)	21.9	(23/105)	9.8	(4/41)	12.5	(1/8)	10.2	(5/49)
5.	"Two Career" problem (balance with spouse's career)	23.9	(16/67)	10.9	(13/119)	20.4	(20/98)	20	(21/105)	9.8	(4/41)	---	---	8.2	(4/49)
6.	Lack of funding/inability to get funding	7.5	(5/67)	4.2	(5/119)	10.2	(10/98)	8.6	(9/105)	4.9	(2/41)	12.5	(1/8)	6.1	(3/49)
7.	Job restrictions (location, salaries, etc.)	9.0	(6/67)	9.2	(11/119)	7.1	(7/98)	5.7	(6/105)	---	---	---	---	---	---
8.	Networking	6.0	(4/67)	<1	(1/119)	0	(0/98)	4.8	(5/105)	2.4	(1/41)	---	---	2.0	(1/49)
9.	Affirmative action	6.0	(4/67)	15.1	(18/119)	14.3	(14/98)	12.4	(13/105)	2.4	(1/41)	---	---	2.0	(1/49)
10.	Positive: active recruitment of women/more opportunities	6.0	(4/67)	10.1	(12/119)	9.2	(9/98)	14.3	(15/105)	14.6	(6/41)	12.5	(1/8)	14.3	(7/49)

11.	Establishing independence	3.0	(2/67)	0	(0/119)	6.1	(6/98)	2.9	(3/105)	---	---	---	---	---	---
12.	Negative social images	3.0	(2/67)	3.4	(4/119)	2.0	(2/98)	<1	(1/105)	2.4	(1/41)	---	---	2.0	(1/49)
13.	Trouble gaining access to nonacademic positions	1.5	(1/67)	1.7	(2/119)	1.0	(1/98)	1.0	(2/105)	---	---	---	---	---	---
14.	Sexual harassment	1.5	(1/67)	<1	(1/119)	2.0	(2/98)	1.9	(2/105)	---	---	---	---	---	---
15.	No answer	0	(0/67)	<1	(1/119)	1.0	(1/98)	1.9	(2/105)	---	---	---	---	---	---
16.	Cut-throat competition	---	---	---	---	1.0	(1/98)	1.9	(2/105)	---	---	12.5	(1/8)	2.0	(1/49)
17.	Gender bias student. evals	---	---	---	---	---	---	---	---	2.4	(1/41)	12.5	(1/8)	4.1	(2/49)

Table II. Categorization of Question 1 Across Year of POWRE Award

Question 1: What are the most significant issues/challenges/opportunities facing women scientists today as they plan their careers?

Categories	*Response Numbers*[b]	*Means of Responses*			
		1997	*1998*	*1999*	*2000*
A Pressures women face in balancing career and family	1, 5, 7	31.9%	30.8%	35.0%	32.4%
B[a] Problems faced by women because of their low numbers and stereotypes held by others regarding gender	3, 4, 8, 10, 12	12.3%	10.1%	9.8%	14.5%
C[a] Issues faced by both men and women scientists and engineers in the current environment of tight resources, which may pose particular difficulties for women	2, 6, 16	10.0%	4.8%	8.2%	7.9%
D More overt discrimination and harassment	9, 11, 13, 14	3.0%	4.4%	5.8%	4.8%

[a]The alphabetic designation for categories B and C have been exchanged, compared with earlier papers (Rosser and Zieseniss, 2000) to present descending response percentages.

[b]Given the responses from all four years, after receiving faculty comments at various presentations of this research, and after working with the data, we exchanged two questions from both category B and D to better reflect the response groupings. Specifically, responses 10 and 12 (considered in category D in Rosser and Zieseniss, 2000) were moved to category B. Similarly, responses 11 and 13 (included in category B in Rosser and Zieseniss, 2000) were placed into category D.

would of male faculty; difficulties in dealing with agencies outside of the university who are used to dealing with male professors; difficulties related to managing demands of scholarship and grantsmanship with maternity demands. More women in a department would possibly allow a better environment for new women faculty members to thrive in such a department through advice/mentoring and more awareness of issues facing women faculty members." (2000 POWRE respondent 26)

> "The biggest challenge that women face in planning a career in science is not being taken seriously. Often women have to go farther, work harder and accomplish more in order to be recognized." (2000 POWRE respondent 21)

> "The CBL Professorship is a tremendous help in two regards. First, simply the prestige of having a named professorship has been useful. Second, the financial security provided by this fellowship has allowed me to undertake risky projects in the lab. Since these are the type of projects that have the highest possible reward, this flexibility is greatly appreciated." (CBL respondent 28)

Category C: Issues Faced by All, with Particular Difficulties for Women

> "I have noticed some problems in particular institutions I have visited (or worked at) where women were scarce. As a single woman, I have sometimes been viewed as "available," rather than as a professional co-worker. That can be really, really irritating. I assume that single men working in a location where male workers are scarce can face similar problems. In physics and astronomy, usually the women are more scarce." (1997 POWRE respondent 26).

> "I still find the strong perception that women should be doing more teaching and service because of the expectation that women are more nurturing. Although research as a priority for women is given a lot of lip service, I've not seen a lot of support for it." (2000 POWRE respondent 1)

> "The fund given in addition to the academic salary has been very useful, especially since the things it could be put toward were left up to us (within reason). I have been able to use this

> fund to start a new project in the lab (that I had not accounted for in my start-up package), hire an undergraduate technician for the summer, and buy computer equipment that made my teaching duties easier." (CBL respondent 4)

Category D: More Overt Discrimination and/or Harassment"

There are almost no women in my field, no senior women, and open harassment and discrimination are very well accepted and have never been discouraged in any instance I am aware of." (1998 POWRE respondent 53)

> "I have often buffered the bad behavior of my colleagues—and over the years I have handled a number of sexual harassment or "hostile supervision" cases where a more senior person (all of them male) was behaving inappropriately toward a lower social status woman (or in rarer cases a gay man)." (1999 POWRE respondent 59)

> "The discrimination they continue to face in the workplace. We seem to be making virtually no gains in terms of rates at which women are granted tenure or promotion to full professor. The older I get, the more depressing these statistics become. Women's research is often marginalized. Women's approaches are not recognized. Men scientists want to judge women by "their" standard (i.e. the white male way of doing things!). Most men have not appreciation for the power and privilege of their whiteness and maleness." (1999 POWRE respondent 70).

Question 2:

The Clare Booth Luce Professors responded similarly (see Table III) to the POWRE awardees to e-mail question 2: "How does the laboratory climate (or its equivalent in your subdiscipline) impact upon the careers of women scientists?" The responses to question 2, in contrast to question 1, reflect less consensus. The response of the CBL professors for "Balancing career and family/time away from home" (response 2) was even stronger (24.5%) than the primary response of the POWRE awardees across all years (15.6 %). Somewhat fewer (2.0%) of the CBL professors than POWRE awardees (12.1%) indicated that they had "not experienced any problems" (response 3). More of the CBL professors than POWRE awardees indicated that they "benefit by working with peers" (response 14), but also that they face a "hostile environment/intimidating/lack of authority" (response 7).

As with question 1, the nuances of difference and context for the responses become clearer from the qualitative answers given by the professors. Many described the impact of negative laboratory climates on the retention of women scientists and the toll these climates take on women's self-esteem:

> "I am fortunate to have worked in laboratories where the environment was very stimulating and supportive. I know many, however, who have had less pleasant experiences. Some of my female peers have left laboratory research altogether because they found the competitiveness of larger laboratories too stressful to cope with." (CBL respondent 31)

> "A practical reality of biochemistry is that, to be highly successful, the scientist must inevitably spend long hours in the lab. This is particularly difficult for women who are trying to juggle small children with work." (CBL respondent 28)

> "The laboratory climate in my field negatively impacts the careers of women scientists. Many of my colleagues are foreign males who do not take females seriously and do not collaborate with them." (2000 POWRE respondent 62).

In contrast, a number of both POWRE and CBL respondents note the efforts that they make to provide a supportive atmosphere in their labs, as exemplified in the following quotation:

> "An open and supportive laboratory climate is very important to the well-being of women scientists. Here at the college, I feel we have a very positive climate for women in our classroom and research laboratories. This is in part due to the high percentage of women in our science classes, reaching almost 70%. A sense of camaraderie often develops. Female students tell me that gender is really a non-issue in the laboratory setting. Doing field work, however, can bring up some gender issues/stereotypes. For example, for some female students it bothers them if male students are stronger and hence do some more of the field work (e.g. pounding in a soil corer more quickly or apparently effortlessly." (CBL respondent 9).

Table III. Impact of Laboratory Climate on Careers

Question 2: How does the laboratory climate (or its equivalent in your subdiscipline) impact upon the careers of women scientists?

	Categories	*1997 POWRE % of responses*		*1998 POWRE % of responses*		*1999 POWRE % of responses*		*2000 POWRE % of responses*		*Current CBL Profs. % of responses*		*Past CBL Profs.% of responses*		*Total CBL Profs. % of responses*	
1.	Don't know/Question unclear	16.4	(11/67)	4.2	(5/119)	7.1	(7/98)	5.7	(6/105)	12.2	(5/41)	12.5	(1/8)	12.2	(6/49)
2.	Balancing career and family/time away from home	13.4	(9/67)	19.3	(23/119)	16.3	(16/98)	13.3	(14/105)	26.8	(11/41)	12.5	(1/8)	24.5	(12/49)
3.	Have not experienced problems	11.9	(8/67)	16.8	(20/119)	10.2	(10/98)	9.5	(10/105)	2.4	(1/41)	---	---	2.0	(1/49)
4.	Not in lab atmosphere/can't answer	11.9	(8/67)	5.9	(7/119)	1.0	(1/98)	8.6	(9/105)	2.4	(1/41)	25.0	(2/8)	6.1	(3/49)
5.	Lack of camaraderie/communications and isolation	9.0	(6/67)	11.8	(14/119)	9.2	(9/98)	14.3	(15/105)	12.2	(5/41)	12.5	(1/8)	12.2	(6/49)
6.	"Boys club" atmosphere	9.0	(6/67)	9.2	(11/119)	18.4	(18/98)	9.5	(10/105)	12.2	(5/41)	---	---	10.2	(5/49)
7.	Hostile environment/intimidating/lack of authority	9.0	(6/67)	14.3	(17/119)	15.3	(15/98)	8.6	(9/105)	19.5	(8/41)	12.5	(1/8)	18.4	(9/49)
8.	Establishing respectability/credibility	9.0	(6/67)	10.9	(13/119)	10.2	(10/98)	3.8	(4/105)	---	---	25.0	(2/8)	4.1	(2/49)
9.	No answer	7.5	(5/67)	6.7	(8/119)	5.1	(5/98)	<1	(1/105)	---	---	---	---	---	---
10.	Positive impact	6.0	(4/67)	10.1	(12/119)	6.1	(6/98)	11.4	(12/105)	2.4	(1/41)	---	---	2.0	(1/49)
11.	Lack of numbering/networking	4.5	(3/67)	6.7	(8/119)	12.2	(12/98)	4.8	(5/105)	2.4	(1/41)	---	---	2.0	(1/49)
12.	General problem with time management	4.5	(3/67)	1.7	(2/119)	5.1	(5/98)	3.8	(4/105)	---	---	25.0	(2/8)	4.1	(2/49)
13.	Safety concerns/presence of toxic substances (health concerns)	3.0	(2/67)	0	(0/119)	4.1	(4/98)	1.9	(2/105)	2.4	(1/41)	---	---	2.0	(1/49)
14.	Benefit by working with peers	3.0	(2/67)	2.5	(3/119)	3.1	(3/98)	5.7	(6/105)	14.6	(6/41)	25	(2/8)	16.3	(8/49)
15.	Problem of wanting research independence	3.0	(2/67)	0	(0/119)	1.0	(1/98)	<1	(1/105)	---	---	---	---	---	---
16.	Lack of funding	1.5	(1/67)	<1	(1/119)	5.1	(5/98)	<1	(1/105)	---	---	---	---	---	---

17.	Benefit from time flexibility/determine own lab hours	3.0	(2/67)	1.7	(2/119)	3.1	(3/98)	1.9	(2/105)	2.4	(1/41)	---	---	2.0	(1/4[illegible])
18.	Did not answer	0	(0/67)	0	(0/119)	3.1	(3/98)	0	(0/105)						
19.	Department doesn't understand basic issues	---	---	---	---	---	---	<1	(1/105)						
20.	Cultural/national stereotypes for women	---	---	---	---	---	---	6.7	(7/105)						
21.	Space	---	---	---	---	1.0	(1/98)	0	(0/105)						
22.	Better bathroom facilities	---	---	---	---	---	---	<1	(1/105)						

"I've built a project and a lab with a group of female scientists. It was a mere coincidence (or was it?) to form an interdisciplinary research visualization group in applied medicine (e.g. virtual surgical training, teaching anatomy via 3D visualization, at [my university's] medical school). Because our group consists of computer scientists, computational linguists, cognitive psychologists, anatomists, we had to establish communication between these disciplines...somehow we managed to develop an amazing climate to collaborate and also attract female graduate students to do research with us." (1998 POWRE respondent 50).

Perhaps the most positive evidence to emerge came from the quotations indicating women's abilities to construct a small, empowering environment within their own labs, within a larger hostile environment:

> "I find the laboratory climate more liberal than, say, the "office climate." I also feel autonomous, powerful and free in this environment (maybe it's because I get to use power tools?) In the laboratory climate, I am able to create and build. I am also able to ask for help and delegate responsibility. Sometimes my colleagues ask me for help. There is a hierarchical structure at the laboratory in which I work, but it is more fluid, roles switch as projects come through. Sometimes I will take the lead and other times I will follow. In terms of my career, working in a laboratory offers a fantastic opportunity to work alone, work with a large group and manage a project, offer support to a colleague, and to build a small community." (1997 POWRE respondent 27).

Concluding Thoughts

Moving from Individual to Systemic Approaches

How can this dilemma faced by academic women scientists and engineers be solved? The 450 women I surveyed are highly educated and successful. They have completed Ph.D. degrees and post-doctoral experiences at the most prestigious institutions in the country. They have succeeded in obtaining a coveted tenure-track position at either a Research I institution or a highly ranked small liberal arts college. Each has competed to obtain a prestigious NSF or CBL award. Most still love their chosen field of science or engineering.

Yet, they express frustration with problems, and in some cases, almost insurmountable barriers erected by institutional and foundational policies and procedures. The interviews and responses to the e-mail questionnaires reveal that some disciplines, institutions, or individual timing of life events are better or worse than others. Encouraging mentors and role models, both male and female, do make a difference. A supportive spouse/partner is critical. But the bottom line remains the same: Most of these women struggle to have both a life and a career as a scientist or engineer.

Systemic Approaches through ADVANCE

Responses to questions 1 and 2 suggest the need for support that extends beyond the research of individual women scientists and engineers. Many of the qualitative statements of awardees from which the categories for the tables emerged particularly underline the need for institutional, systemic approaches to balance career with family, deal with problems resulting from low numbers of women in some disciplines and the stereotyping they may encounter, as well as more overt discrimination and harassment. (*7, 8*).

The relatively new ADVANCE program (institutional transformation component) at the National Science Foundation funded nine universities beginning in FY 2001 (*15*) and funded a similar number in FY 2003 to develop model policies and practices to address institutional barriers and discouragements faced by female science, technology, engineering, and mathematics faculty. The results of ADVANCE will provide a variety of models for improving the environment in academic science and engineering departments and transform faculty careers to be more attractive and supportive of all men and women, particularly those from previously underrepresented populations.

The study described in this chapter, comparing the responses of POWRE awardees (most at research institutions) with those of CBL Professorship recipients (many at small liberal arts colleges and faith-based institutions) helps in understanding the experiences of women faculty and the barriers they face across a broad spectrum of academic settings. This understanding points toward institutional policies or practices which could increase the satisfaction, retention, and success of female faculty in fields where they are least well-represented. In the following chapter, CBL Program Officer Jane Daniels describes how policy changes might enhance the promotion and retention of women faculty. Such positive changes should have a ripple effect on female graduate and undergraduate students as they consider the wisdom of choosing a career in academia.

References

1. Hopkins, N. MIT and gender bias: Following up on victory. *Chronicle of Higher Education* **1999**,. (December 3), B-4.
2. Campbell, K. Leaders of 9 universities and 25 women faculty meet at MIT, agree to equity reviews. MIT News Office **2001**, Retrieved January 31, 2001 from www.mit.edu/newsoffice/nr/2001/gender.html.
3. National Science Foundation *Professional opportunities for women in research and education Program solicitation.* **1997**, 87-91. Arlington, VA:
4. Evangelauf, J. A new Carnegie classification. *The Chronicle of Higher Education*, **1994,** (April 6).
5. Rosser, S. V Balancing: Survey of fiscal year 1997, 1998, and 1999 POWRE Awardees. *Journal of Women and Minorities in Science and Engineering.* **2001**, *7*(1), 1-11.
6. Rosser, Sue V. and Lane, Eliesh O'Neil.. A history of funding for women's programs at the National Science Foundation: From individual POWRE approaches to the ADVANCE of institutional approaches. *Journal of Women and Minorities in Science and Engineering,* **2002**, *8*(3-4), 327-346.
7. Rosser, S. V. and Lane, E. O'Neil. Key barriers for academic institutions seeking to retain women scientists and engineers: Family unfriendly policies, low numbers, stereotypes, and harassment. *Journal of Women and Minorities in Science and Engineering* **2002**,. *8*(2), 163-191.
8. Rosser, S. V. and Zieseniss, M. Career issues and laboratory climates: Different challenges and opportunities for women engineers and scientists (Survey of fiscal year 1997 POWRE awardees*). Journal of Women and Minorities in Science and Engineering,* **2000**, *6*(2), 1-20.
9. Henry Luce Foundation. *The Clare Boothe Luce program for women in science, mathemactics and engineering.* **2000**, New York: .
10. Schneider, A. Female scientists turn their back on jobs at research universities. *The Chronicle of Higher Education* **2000**,. (August 18), A12-A14.
11. Kuh, C. You've come a long way: Data on women doctoral scientists and engineers in research universities. In L. Hornig (Ed), *Equal rites, unequal outcomes*. New York: Kluwer Academic, 2000, 111-144..
12. Curry, D. Prime numbers. *The Chronicle of Higher Education,* **2000**,. (July 6), A9.
13. National Science Foundation. *Women, minorities and persons with disabilities in science and engineering*. Arlington, VA, **2000**, (NSF 00-327).
14. Rosser, S. Different laboratory/work climates: Impacts upon women in the workplace. In C. Selby (Ed.), *Women in science and engineering: Choices for success.* New York: Annals of the New York Academy of Sciences, 2000, 95-101.
15. National Science Foundation.. *ADVANCE Institutional Transformation* http://www.nsf.gov/advance. Retrieved October 1, 2001.

Chapter 6

The Clare Boothe Luce Program for Women in the Sciences and Engineering

Jane Zimmer Daniels

The Henry Luce Foundation, 111 West 50th Street, Suite 4601, New York, NY 10020

The Clare Boothe Luce Program for Women in the Sciences and Engineering is the largest private source of support for women in the sciences and engineering, with grants totaling over $100 million to support 1430 scholarships, fellowships and professorships since its inception in 1989. A group of the professorship recipients participated in a study comparing their responses with those of NSF-funded POWRE awardees. The experiences of these pre-tenure female faculty members across a broad spectrum of academic settings suggest institutional policies or practices which would increase the satisfaction, retention, and success of female faculty members in fields where they are least well-represented. The understandings provided by this research on women have significant potential to enhance the career development, work environment and retention of men as well.

The Clare Boothe Luce Program

The Clare Boothe Luce (CBL) Program strives to increase the participation of women in the sciences (including mathematics) and engineering at every level of higher education and to serve as a catalyst for colleges and universities to be proactive in their own efforts toward this goal. The CBL Program is the single largest private source of funding for women in science and engineering. Since its inception grants totaling over $100 million have been made to 137 different institutions supporting 1430 women with scholarships, graduate fellowships or professorships. Figure 1 shows the total number of recipients supported and the total expenditures for each grant category.

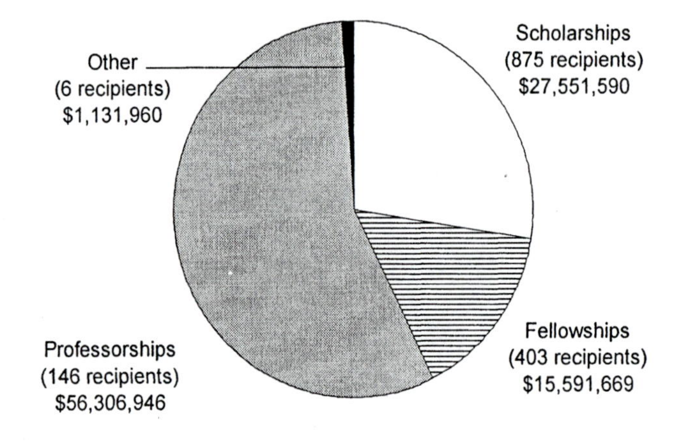

Figure 1. Clare Boothe Luce Program Grants 1989-2004

Clare Boothe Luce (1903-1987) was a remarkable woman whose career spanned seven decades and nearly as many professional interests, such as journalism, politics, the theatre, diplomacy, and military intelligence. In each of those fields she excelled. Not content with her achievements, Mrs. Luce was always eager to consider new topics, to test new hypotheses, and to encourage other women to achieve their own potential. Characteristically, she declined to restrict her vision to the fields in which she had established her reputation. Under the terms of her will, she chose instead to establish a legacy that would benefit current and future generations of women with talent and ambition in areas where they continue to be severely underrepresented—the sciences and engineering.

In establishing the Program, Mrs. Luce designated as its administrator, The Henry Luce Foundation, established in 1936 by her husband, Henry R. Luce, co-founder and editor in chief of Time, Inc. Grants are made in one of three categories: 1) undergraduate scholarships, 2) graduate fellowships, and 3) term support for beginning tenure-track appointments. As stated in her will the program is intended "to encourage women to enter, study, graduate, and teach" in fields where there have seemingly been obstacles to their advancement. All physical and life sciences, mathematics, computer science and all areas of engineering are included. Medical sciences are excluded.

Clare Boothe Luce Professorships

Grants for Clare Boothe Luce (CBL) Professorships are made to colleges or universities for a new tenure-track faculty position. Once the grant is received, the institution has a year to search for an appropriate female candidate for the CBL Professorship. The grant covers the recipient's salary, benefits and a career development fund for a period of five years. The institution must show evidence of their ability to support the position after the grant period. The intent of the professorship is to identify female scientists and engineers of the highest caliber and to guarantee early in their academic career, opportunities commensurate with their considerable talents. The candidate must be external to the institution's existing faculty, typically in her first tenure-track position. Any candidate for the professorship must be either a U.S. citizen or permanent resident. The recipient is identified as a Clare Boothe Luce Assistant (Associate) Professor. Figure 2 shows the breadth of disciplines represented by the CBL Professorship recipients.

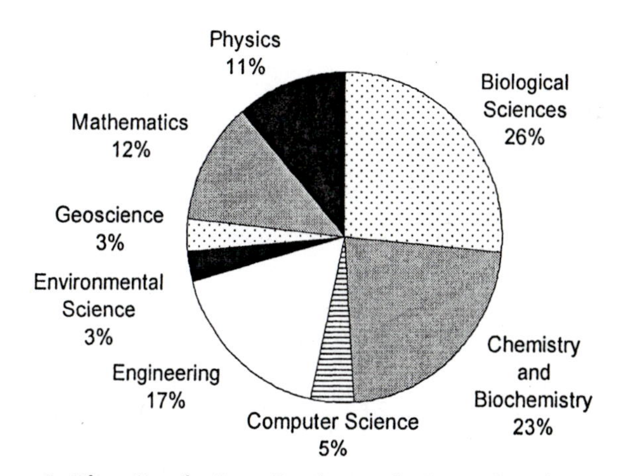

Figure 2. Clare Boothe Luce Professors by Discipline (1989-2004)

A successful proposal typically describes how the institution plans to increase the recipient's external visibility and nurture her professional development (e.g. mentoring by senior faculty members, resources for research, additional travel funds, relief from administrative duties). The proposal also indicates the administration understands the factors that may hinder a woman's career advancement and describes how the ***institution*** provides support for all female faculty members in the sciences and engineering to ensure their success. Quantifiable evidence of such support (e.g. comparable tenure and promotion rates for female and male faculty members, prior successful assimilation of female faculty members into a nationally recognized research program, existing career development programs for faculty) is requested.

An identifying characteristic of the CBL Professorship grant is a special career development fund (typically 20% of salary). The unusual feature of this fund is its great flexibility. Not only may it be used for typical research expenses, such as equipment and graduate assistants, but for release time, travel, and child-care. The recipient of the professorship acts as the principal investigator for this allocation and is mentored by an experienced faculty member to help the CBL Professor leverage these funds in the most effective ways. This allocation is in addition to normal start-up funds provided by the institution. The institution must provide the facilities and resources required by the recipient of a value equal to or greater than those provided to comparable faculty members.

Policy Implications of CBL Professors' Experiences

The research described by Rosser in Chapter 6 of this volume (*1, 2, 3*) compared the responses of POWRE awardees (most at research institutions) with those of CBL Professorship recipients (many at small liberal arts colleges and faith-based institutions). POWRE awardees are women who received peer-reviewed funding from a focused National Science Foundation program in fiscal years 1997-2000. They are primarily untenured assistant professors in tenure-track positions at research universities. The POWRE awards were capped at $75,000, with a typical duration of 12-18 months.

The experiences of the POWRE and CBL pre-tenure female faculty members across a broad spectrum of academic settings points toward institutional policies or practices which could increase the satisfaction, retention, and success of female faculty members in fields where they are least well-represented. Such positive changes have a potential ripple effect on female graduate and undergraduate students as they consider the wisdom of choosing a career in academia.

The issue of balance—whether pertaining to the tension between personal responsibilities and the demands of work or among competing demands within the work environment—surfaces time and again as an impediment to the attraction and advancement of women in the sciences and engineering. The combined responses of POWRE awardees and CBL Professorship recipients leave no doubt that the issue of balancing work with personal responsibilities is the most pervasive and persistent challenge facing female science and engineering faculty members—irrespective of the type of institution or discipline. The conflicting demands of work and personal responsibilities are likely exacerbated for female science and engineering faculty members because of the competitiveness and inflexibility characteristic of these fields. Engineering in particular, with its origins in the military (*4*; *5*), unnecessarily perpetuates its hierarchical nature and cutthroat competitiveness.

Many female faculty members, if they have postponed childbearing until after graduate school and post-doctoral experiences, face a common dilemma—how to resolve the competition between the biological clock and tenure clock. An additional challenge makes this situation more problematic for female faculty members in the sciences and engineering since most of them (62%) are married to a scientist or engineer who has similar, unreasonable demands on his time (*6*). Although most of their male colleagues are also married, few are married to a scientist or engineer.

Increasing Flexibility and Distributing Control

Creative institutional solutions to faculty members' pressures to balance work and personal demands appear promising. Such solutions cluster around two issues—increased flexibility for individual faculty members and the distribution of control from the institution or administration to the individual. Flexibility evident in work hours, benefits, and telecommuting and distribution of control to the individual through a cafeteria system of benefits or in a start-up package that includes a professional development account available until tenure (hallmark of the CBL Professorships) are examples of such institutional solutions.

Several institutions have developed a cafeteria of benefits that provides important flexibility across the span of a faculty member's career. Child care or elder care benefits may take the form of financial assistance, information and assessment of available services, or the convenience of on-site facilities. An example of this type of flexibility exists at Iowa State University where on-site child care for infants through kindergartners has a sliding scale of fees; a cafeteria of benefits allows any benefit dollars that remain after selection to be moved to a flexible spending account for medical expenses or child care expenses, departmental assistance for spousal hires, and lactation rooms for nursing women.

Institutional policies that address the issue of balance would likely have a positive impact on the recruitment and retention of female undergraduate and graduate students, as well. The perception that success in the sciences and engineering requires an intense, unbalanced focus on inanimate objects for prolonged periods of time is a significant deterrent to women selecting those fields for a life-long career (*7*).

Broadening Access to Success

The small number of women in most areas of the sciences and engineering can impede, or worse yet, end, the career of an outstanding female scientist or engineer. A lack of role models, feelings of isolation, and stereotyping are all obstacles caused by low numbers. In recent years interventions at the pre-college, undergraduate and graduate levels have resulted in small increases in enrollment and graduation rates of females. A number of institutions are now broadening their scope to include faculty programs intended to offset the consequences of the small proportions of women in their ranks. Examples of such faculty programs include mentoring of new faculty members by senior faculty members; networking events for female faculty members to diminish feelings of isolation; and structuring of departmental/college symposia to ensure the inclusion of distinguished female speakers.

The initial steps of widening paths to success in academia require an understanding of gender differences specific to an individual institution or department. Do paths narrow in certain places (departments), at critical junctures (recruitment, tenure, promotion to full professor, prestigious awards, or influential committees) or over specific issues (salary, space, or graduate student assignment)? A study of faculty members hired as assistant professors or instructors at the University of Michigan between 1982 and 1988 revealed that 53% of male assistant professors and 24% of male instructors but only 43% of the female assistant professors and 10% of the female instructors received tenure (*8*). Surveys such as the ones conducted at MIT (*9*) identified sources of inequity that further restrict the already narrow paths to successful advancement and recognition for female faculty members. Sources of inequity identified in the MIT survey results included such things as women's exclusion from group grants, women not being invited to serve on the PhD thesis committees of the students of male colleagues, and women feeling a lack of influence in important departmental decision making.

Creating additional, equally valued paths to success also widens career opportunities. Alternatives to traditional procedures for advancement and recognition that hold promise for greater effectiveness with future faculty members might be discovered through a brainstorming session or survey of beginning faculty members and graduate students. When alternate paths to success are created, they must be viewed as equally prestigious and attractive to faculty members regardless of gender, age, race, or ethnicity in order to produce positive results.

Another group of barriers to the successful career advancement of women in academia are found in advancement and recognition practices. General requirements for achieving tenure and promotion are typically provided to both men and women, however, informal avenues of communication often result in inequitable results. For example, new faculty members are usually told the number of articles they must publish in refereed journals in order to achieve tenure in a particular department. Months or years later in an informal setting, a full professor may tell his male colleague that the paper he is developing should be published in a specific journal to result in the greatest approbation of senior faculty members in that department. In addition, the full professor may offer to introduce his less experienced protégé to the editor of that journal at the next professional meeting. Such information is not intentionally withheld from female faculty members; it simply doesn't get communicated as frequently. Providing more structure and transparency to the advancement and recognition practices in individual departments also widens paths. Negative forms of discrimination are less likely to occur if the paths to academic advancement and recognition are clearly understood by both the beginning faculty members who must negotiate them and the senior faculty members responsible for their implementation (*10*) Examples of such transparency include a panel of newly tenured faculty members speaking to new faculty members or an effective third-year review process that identifies potential weaknesses in an untenured faculty member and provides a plan for addressing those weaknesses.

Making a Positive Impact on the Environment

Female faculty members in science and engineering departments often describe their work environment with words such as chilly, masculine, exclusionary, elitist, and hostile. The differences between the availability of female Ph.D.s and the actual proportion of women on faculties vary considerably by discipline, with especially large discrepancies in chemistry and mathematics (*11*) In chemistry, where the proportion of women completing Ph.D.s has been above 20% since 1985 (*12*), the proportion of those choosing to return to the inhospitable environments that educated them is closer to 12% at the top 50 universities (*13*). An in-depth interview study with female faculty members who left the University of Michigan "voluntarily" uncovered the extent to which the women "cited lack of respect by their colleagues as figuring in their decision" to leave the institution (*14*).

Lack of collegiality and difficulty in gaining credibility among their peers in science and engineering departments exemplify characteristics of a negative environment identified by the research on POWRE awardees. This issue was not identified as often among CBL professorship recipients, perhaps due to the more prestigious nature (named professorship, larger award, and longer duration) of the grant. POWRE awardees found that their peers viewed a "women's award" as less prestigious. In fact, these awards were smaller in size and less competitive that most NSF research awards. The CBL professorship recipients commented on the opposite effect. A "named" professorship that provided full salary support for five years, augmented by a sizable and very flexible career development fund, enhanced the woman's credibility, and was perceived by peers and senior faculty members as prestigious, provided experience administering a research account, and set high expectations for excellence among the recipient's peers.

Wadsworth (*15*) suggests other ways of improving the environment in *Giving Much, Gaining More*. The book describes the personal impact of mentoring programs developed at Purdue University in the 1990s that successfully used positive actions to offset the negative characteristics of engineering departments—welcoming vs. excluding, communicating vs. bickering, trusting vs. doubting, accepting vs. rejecting, and affirming vs. ridiculing. If such positive actions became the "norm" in science and engineering departments, the need for such supplemental, support programs for women would eventually disappear.

In FY2001 and FY2002 the National Science Foundation's ADVANCE program (Institutional Transformation component) funded eighteen colleges and universities to develop model policies and practices that address institutional barriers and discouragements faced by female science, technology, engineering, and mathematics faculty members (*16*). The ADVANCE institutions are providing a variety of models for improving the environment in academic science and engineering departments and transforming academic careers in ways that are more attractive and supportive of all men and women, particularly those from previously underrepresented populations. Results showing tools, programs, policies and practices are posted to the project websites of each of the ADVANCE Institutional Transformation sites and can be accessed at http://research.cs.vt.edu/advance/tiki/tiki-index.php?page=AdvanceInstitutions or the websites for individual institutions: University of Alabama at Birmingham; University of California, Irvine; Case Western Reserve University; University of Colorado, Boulder; Georgia Institute of Technology; Kansas State University; University of Maryland, Baltimore County; University of Michigan, Ann Arbor; University of Montana; New Mexico State University; Hunter College, City University of New York; University of Puerto Rico, Humacao; University of Rhode Island; University of Texas at El Paso; University of

Washington; University of Wisconsin, Madison; Utah State University; Virginia Polytechnic Institute and State University.

References

1. Rosser, S.V. Balancing: Survey of Fiscal Year 1997, 1998, and 1999 POWRE Awardees. *Journal of Women and Minorities in Science and Engineering.* 20004, vol. 6, pp. 1-11.
2. Rosser, S.V.; Daniels, J.Z. Widening Paths to Success, Improving the Environment, and Moving Toward Lessons Learned from the Experiences of POWRE and CBL Awardees. *Journal of Women and Minorities in Science and Engineering.* 2004, vol. 10, pp. 131-148.
3. Daniels, J.Z.; Rosser, S.V. Examining the Problem of Underrepresentation Through a Study of Award-Winning Faculty. AWIS Magazine. 2004, vol. 32, no. 3, pp. 12-21
4. Hacker, S. *Pleasure, Power and Technology*; Unwin Hyman: Boston, MA, 1989.
5. Cockburn, C. The Material of Male Power. *Feminist Review.* 1981, vol. 9, pp. 41-58.
6. Sonnert, G. and Holton, G. *Who Succeeds in Science? The Gender Dimension.* Rutgers University Press: New Brunswick, NJ, 1995.
7. Margolis, J. & Fisher, A. *Unlocking the Clubhouse*; The MIT Press: Cambridge, MA, 2002; pp 72-73
8. Hollenshead, C. Women in the academy: Confronting Barriers to Equality. In L. Hornig. *Equal Rites, Unequal Outcomes*. Kluwer Academic: New York, NY, 2003; pp. 211-225.
9. Hopkins, N.; Bailyn, L; Gibson, L.; and Hammonds, E. *The Status of Women Faculty at MIT.* [http://web.mit.edu/faculty/reports/overview.html], 2002.
10. Fox, M. F. Women and Scientific Careers in S. Jasanoff, J. Marble, J. Petersen, & T. Pinch (Eds.) *Handbook of Science and Technology Studies.* Sage: Newbury Park, CA, 1995; pp. 205-233.
11. Hornig, L. (2003). The Current Status of Women in Research Universities. In L. Hornig. *Equal Rites, Unequal Outcomes*. Kluwer Academic: New York, NY, 2003; pp. 31-51.
12. American Chemical Society. Percentage of Chemistry Degrees Earned by Women from 1967 to 1999. *ACS Starting Salary Survey, 1999.*
13. *Chemical and Engineering News*. September, 2002; pp. 110-111.

14. Hollenshead, C. Women in the academy: Confronting Barriers to Equality. In L. Hornig. *Equal Rites, Unequal Outcomes*. Kluwer Academic: New York, NY, 2003; pp. 211-225. , p. 219.
15. Wadsworth, E. M. *Giving Much Gaining More*; Purdue University Press: West Lafayette, IN, 2002.
16. National Science Foundation. *ADVANCE Institutional Transformation* http://www.nsf.gov/advance. Retrieved October 1, 2001.

Chapter 7

Equal Opportunity in Chemistry in Germany

Sonja M. Schwarzl

Computational Molecular Biophysics, University of Heidelberg, Im Neuenheimer Feld 368, 69120 Heidelberg, Germany (email: sonja.schwarzl@iwr.uni-heidelberg.de)

Gender equality has been a recurrent theme for the past decades. This article reviews the current status and, where available, the development over the past 20 years in chemistry in Germany. Statistical data are presented that summarize the situation in both academia and industry. Possible reasons for the observed gender disparities are outlined and recommendations for future activities are given.

Although gender equality is a recurrent theme in the public discussion, equal opportunity for women and men are far from being realized on a large scale. In Germany the public discussion on this issue in the field of chemistry started later than in the other fields of science and engineering. This is reflected in the fact that the Arbeitskreis Chancengleichheit in der Chemie (AKCC) (*1*), founded in March 2000 (*2*) as a division of the German Chemical Society, is younger than its sister organizations. For comparison, it should be mentioned that the division Frauen im Ingenieurberuf (*3*) (women engineers) of the Verein Deutscher Ingenieure, the engineering equivalent to the German Chemical Society, was founded in 1982 (*4*), the Deutsche Ingenieurinnen Bund (*5*) (German Women Engineers' Society) in 1986, and the Arbeitskreis Chancengleichheit (*6*) of the German Physical Society in 1998 (*7*). The general discussion shows that the basic observations regarding the situation for women and men are similar, irrespective of the specific scientific discipline within Germany as well as in international comparison. Naturally, discussions about the range or the reasons for the observed inequalities are frequently biased, because the individuals are personally involved. Often the facts remain unclear, because general trends are being constructed from individual experience or anecdotal evidence without making use of available data from statistical surveys. Such data are scarce, however, as stated by the European Technology Assessment Network in 2002: "The major difficulty in reviewing the position of women in science in the EU today is the lack of systematically collected and published statistics." (*8*)

Statistical data for chemistry in Germany are available from the annual surveys of the German Chemical Society (*9*) and the Statistisches Bundesamt (*10*), as well as surveys among the members of the Verband angestellter Akademiker und leitender Angestellter der chemischen Industrie (*11*) and additional surveys. The current situation is reviewed and commented on based on these statistical data.

Statistics

University

Sources for gender-disaggregated statistics from the chemistry departments of German universities are available through the annual surveys of the German Chemical Society and the Statistisches Bundesamt. Figure 1 shows the decrease in percentage of women with qualification level throughout the course of university education as of 2003. Data for the habilitation are for 2002. In 2003, almost 50% of beginners were female through the levels of Vordiplom (roughly corresponding to a Bachelor's degree) and Diplom (roughly corresponding to a Master's degree). However, the percentage of women drops to about 25% at the level of Ph.D. and less than 20% at the level of habilitation. In the figure, both actual and potential percentages of women and men are given. The actual percentages are those statistically observed. The potential percentages correspond to the values one would expect if men and women at a given qualification level had the same success rate in reaching the next higher qualification level. The potential percentages were determined considering the average qualification time for each level. The potential percentage at the beginners' level is identical to the percentages of women and men among high-school graduates of the same year. At the levels of Vordiplom, Diplom, and Ph.D. the potential percentages are taken from the actual percentages at the beginners' level two, six, and ten years earlier. At the habilitation level, the potential percentages are taken from the actual percentages at the level of Ph.D. six years earlier.

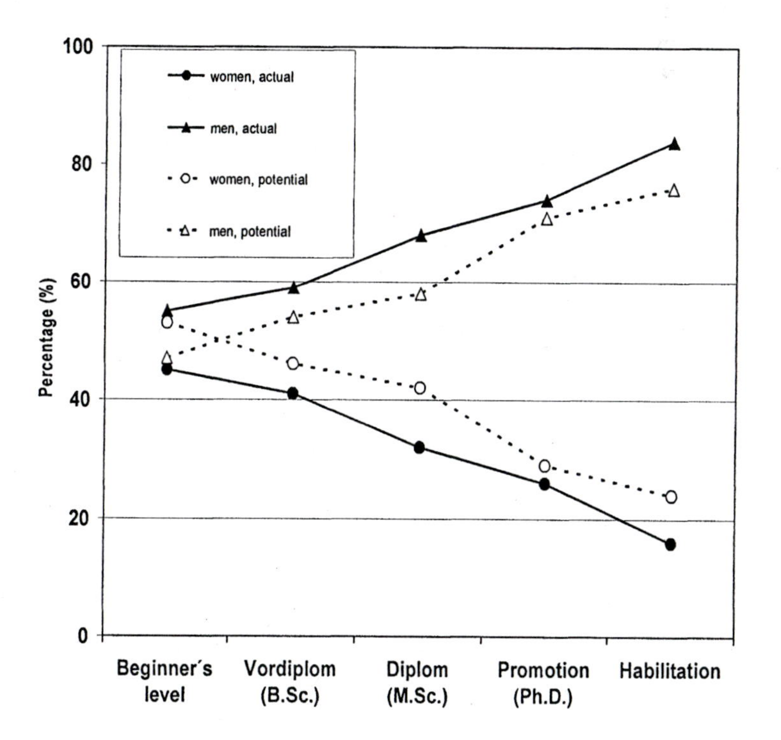

Figure 1: Actual and Potential Percentages of Women in Chemistry during the Qualification Phase in 2003.
SOURCES: GDCh and Statistisches Bundesamt

The actual percentages of women are always lower than the potential percentages, a phenomenon known as the Leaky Pipeline for women. Strikingly, the actual percentages are almost parallel to the potential percentages. This indicates that women who have decided to study chemistry and who have confirmed their decision by passing the Vordiplom exam are likely to successfully finish their studies and continue their studies up to the qualification stage of Ph.D. A similar conclusion has been drawn recently from a comparison of success rates of women and men at the levels of Vordiplom, Diplom, and Ph.D. for beginning students of the years 1992, 1995, and 1997 (*12*).

Figure 2 shows a time course of the percentage of women and men from 1984 to 2003 in the Qualification Stage. Already in 1984, 31% of beginners were female. This percentage remained almost constant up to 1994. Starting in 1995 the percentage of women beginners rose significantly and reached almost 50% in 1999 and has remained at that level. This illustrates that young women are equally interested in taking up chemistry at universities as young men. Remarkably, the rise in percentage of women coincided with a dramatic drop in the absolute number of beginners (Figure 3). In 1994 and 1995 in Germany, fewer individuals started studying chemistry than those that earned a Ph.D. in chemistry.

The extreme drop in enrolment resulted in the implementation of several programs to encourage young people to study chemistry. Examples are Laboratories for Pupils (*13, 14*), the Bildungsinitiative Chemie (*15*), the development of advanced training centers for teachers (*16,17*), scholarships for undergraduates (*18*) by the Verband der Chemischen Industrie (*19*), partnerships with schools (*20*), and mentoring projects such as the Ada-Lovelace-Project (*21*). Some of those programs explicitly focus on encouraging young women.
One can therefore assume that the rise in the percentage of women among beginners stems at least partly from those efforts. It is moreover remarkable that the percentage of women among beginners has stayed high in recent years, while the absolute number of beginners continued to increase. Whether there will be an equal distribution of the genders at the higher qualification levels in the future is an open question. However, the data shows unambiguously that it takes longer to reach an equal distribution on a given qualification level than the average qualification time. For example, it takes about two years to reach the Vordiplom level. In 1995, 36% of beginners were female. An approximate equivalent percentage of 33% was reached only in 1999 for the level of Vordiplom. Thus, it took four years or double the qualification time until the rise in percentage of women among the beginners propagated to the next higher qualification level.

In contrast to the dynamic gender distribution of the undergraduate and graduate students, the percentage of women on university staffs has only changed marginally in recent years. Figure 4 shows a time course of the percentages of female and male faculty members over the past 20 years. Until

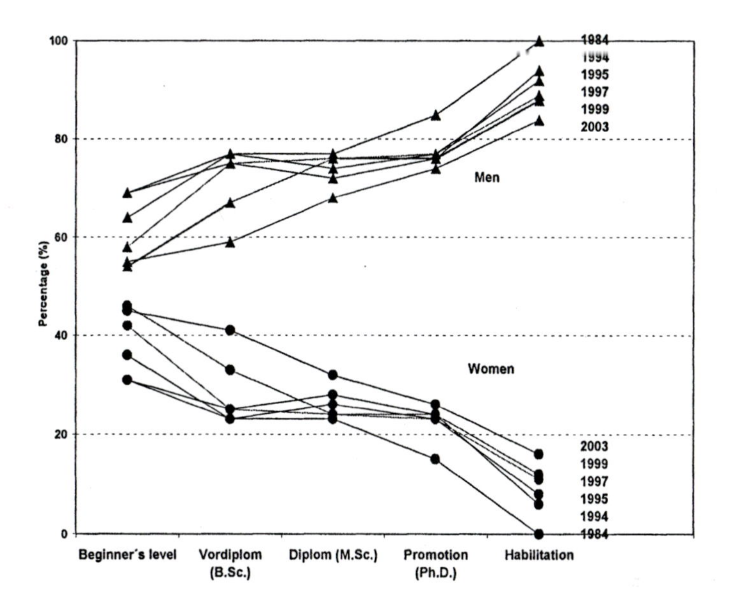

Figure 2: Time Course of the Percentages of Women and Men in the Qualification Phase in Chemistry from 1984 to 2003. Data for the habilitation are for 2002. Sources: GDCh and Statistisches Bundesamt

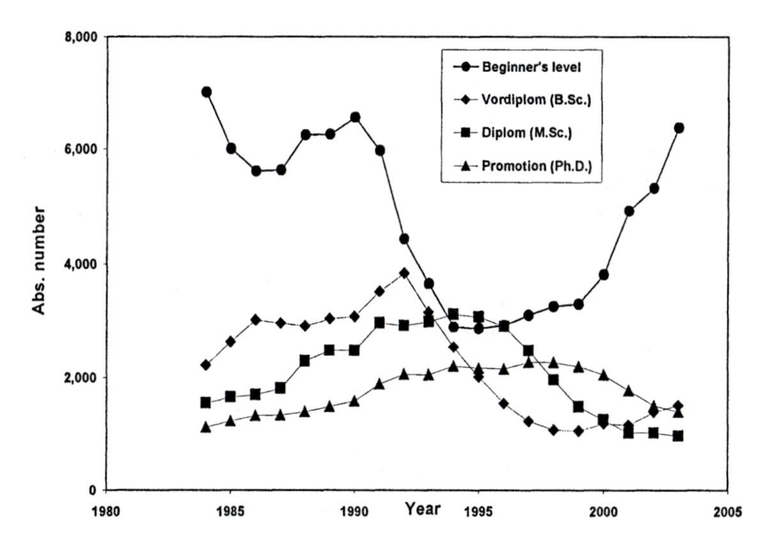

Figure 3: Absolute Number of Beginners and Graduates in Chemistry from 1984 to 2003 (Source: GDCh).

1997, the percentage of women among full professors (C4 and C3) was below 3%. In the past five years this percentage has increased to 4.0% at the C4 level and 6.7% at the C3 level. These percentages are shockingly low compared to the rather high number of women among the students. At the C2 level (roughly comparable to associate professor level) a more profound rise in percentage of women has occurred. However, this rise is insignificant since these positions are currently being abolished due to the restructuring of the German university laws and the absolute number of C2 positions is constantly decreasing (see Figure 5).

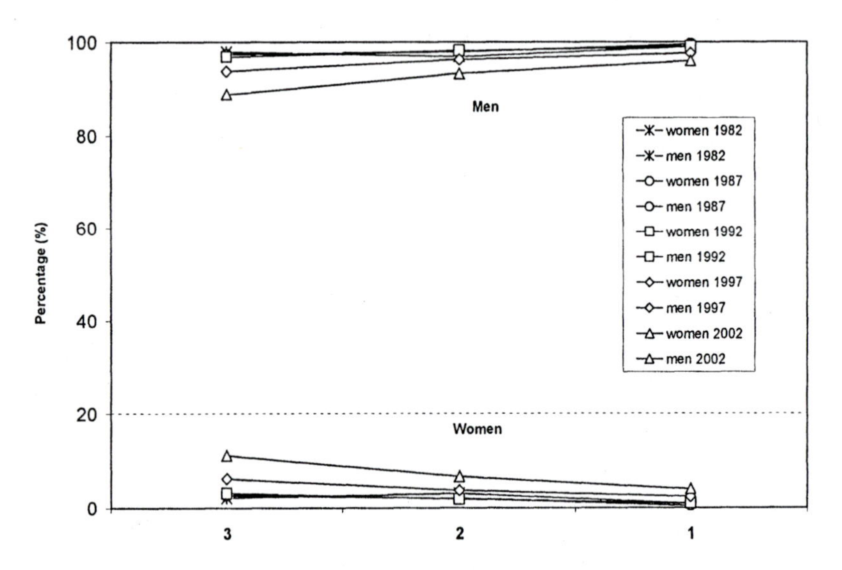

Figure 4: Time Course of the Percentages of Female and Male University Staff in Chemistry from 1984 to 2002. Categories 1, 2, and 3 Correspond to C4, C3, and C2 Positions, respectively. Source: Statistisches Bundesamt.

Although as many as 15% of all Ph.D.s in chemistry were awarded to women in 1984, they are not represented in the same proportion at the professor levels 20 years later. This trend is observed throughout all European countries. "The percentage of women among university professors does not even reach 20% in any country of the European Union – it is thus below the percentage of women Ph.D.s reached already 20 years ago. There is no lack in adolescent female talents in science." (*22*)

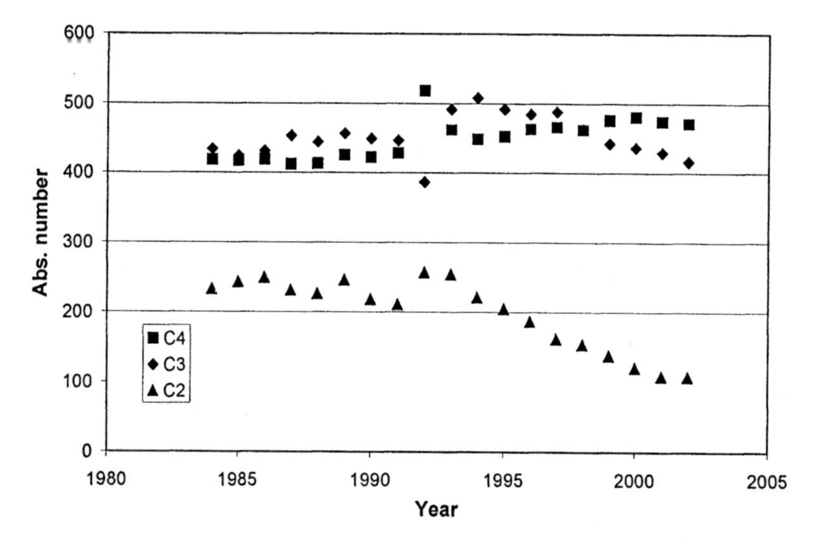

Figure 5: Total Numbers of University Staff in Chemistry from 1984 to 2002. Source: Statistisches Bundesamt.

A comparison across disciplines within Germany made by the Bund-Länder-Kommission states that the trend is not restricted to chemistry or other natural sciences. "Career breaks occur especially between graduation and Ph.D. and between Ph.D. and habilitation. This leads to enormous under-representation of women in leading positions." (*23*) The factors leading to this plainly visible selection process and whether there is a special environment influencing the situation in chemistry has been discussed in detail using the example of the chemistry faculty at the University of Göttingen (*24*). In this study it was found that the assessment of the suitability of applicants by decision makers played an important role at all stages, *i.e.*, equally affecting applicants for diploma, Ph.D., and habilitation positions and applicants for tenure. In the course of this assessment they detected that the observed, but not explicitly outspoken mental construction of a successful chemist to be male resulted in the exclusion of women, without being obviously discriminatory.

Industry

Gender-disaggregated statistical data concerning the German chemical industry are not easily accessible. However, the Verband angestellter Akademiker und leitender Angestellter der chemischen Industrie conducts representative surveys on a regular basis for the leaders and decision makers in the German chemical industry. Two surveys in 1994 and 2000 focused on equal opportunity for both genders. The next survey of this type is scheduled for the fall of 2005. The results of the 1994 (*25*) survey are shown in Figure 6. Twelve percent of women self-reported that they were at clerk/general-employee level, 3.2% on the level of group leader, 2% at the level of department head and 0% at the board level. This situation remained unchanged up to the year 2000 (*26*). Only in September 2004 was the first woman appointed as board member in a German chemical company.

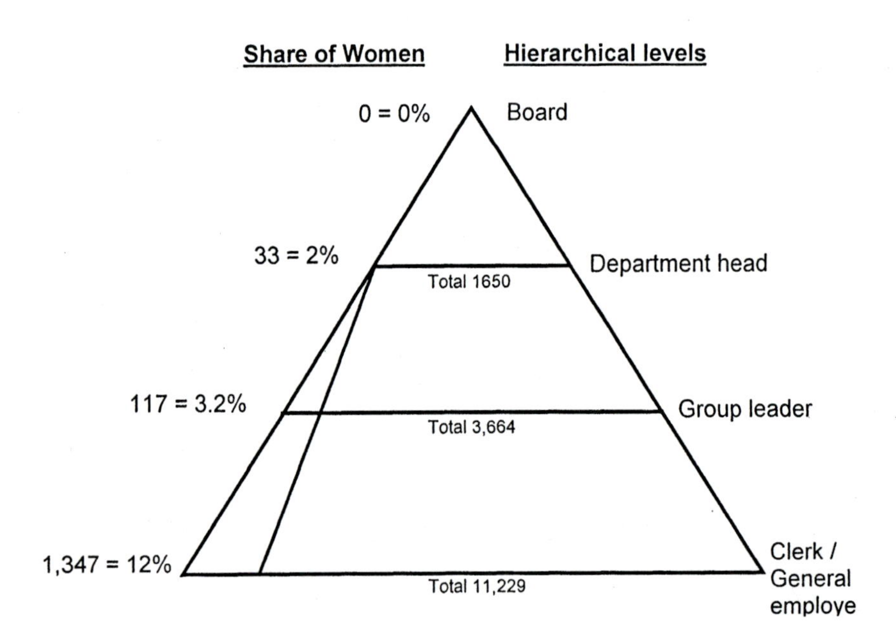

Figure 6: Percentages of Women and Men in German Chemical Industry in 1994. The Figure was Adapted from Ref. 25. In the Original German Legend the Hierarchical Levels are Termed Vorstand, Geschäftsführung; Bereichsleiter/in, Abteilungsleiter/in; Betriebsleiter/in, Gruppenleiter/in; Sachbearbeiter/in mit und ohne Mitarbeiter.

In comparison to other European countries, Germany has the lowest percentage of women working in industrial research. This was documented impressively in the report to the European Commission "Women in Industrial Research. A Wake Up Call for European Industry" (*27*). Table I shows the total number of industrial researchers in the European Union countries and the percentage of female workers. With about 150,000 researchers, Germany has the largest workforce, and along with Austria it has the lowest percentage of female industrial researchers (9.6%). In order for the percentage of female industrial researchers to rise in Europe, there must be an increase in the number of female researchers in Germany. For this to occur, the percentage of women working in chemical industries will have to be increased, since the chemical industry is a large portion of industrial research in Germany.

Table I: Number and Percentage of Female Industrial Researchers Working in the European Union Countries in 1999.

Country	*All researchers*	*Number of women*	*% of women*
Germany*	150,149	14,414	9.6
Denmark	11,292	2,218	19.6
Greece	3,931	940	23.9
Spain	17,310	3,353	19.4
France	86,215	17,787	20.6
Ireland	1,900	536	28.2
Italy	29,706	5,490	18.5
Luxembourg**	1,217	No data	No data
Austria	13,966	1,258	9.0
Portugal	3,328	793	23.8
Finland	22,515	3,999	17.8
Sweden**	39,921	No data	No data
UK**	98,587	No data	No data
EU (10)	340,312	50,789	14.9

Source: DG Research, Unit C5: Data: Eurostat, New Cronos; DG Research, WiS database. The table is reproduced from Ref. 27.

Exceptions to the reference year: Austria (1998); France, Italy (2000); Ireland (2001)

* = Full time equivalent; ** = No gender differentiation data available

No data for Belgium and the Netherlands

General Situation

Currently, the "structural barriers for graduates from technical and natural scientific disciplines" are being studied by the Institute of Sociology of the Technical University of Darmstadt (Prof. Beate Krais). In chemistry, about 7500 questionnaires were sent to all female members of the German Chemical Society and an equivalent number of men were polled. The initial findings have been published (*28*) and are similar to a previous study that has investigated the field of physics (*29*). The average annual income of female chemists is significantly lower than that of male chemists. However, one has to take into consideration that the women who were surveyed were younger and would be expected to hold lower positions having less pay than the men.

As expected, the study shows that for a given age group, men held higher positions than women. For example, only 10% of the women but 25% of the men above the age of 45 held positions as head of a division or higher. Both women and men evaluated the working climate and their relationship with their supervisors as good, with slight gender-specific differences in favor of the men. Men reported attending significantly more personal development courses and seminars for higher management than the women. This is in sharp contrast to the fact that both women and men equally expressed an interest in developing their skills. As the age of the respondents increased, this imbalance in treatment between the men and women grew.

Visible differences between the genders could also be found in their private lives. Female chemists lived more often without a partner than male chemists. If the women lived with a partner, they more often maintained separate households. This mirrors the choice of partners: 80% of the affiliated females had partners having advanced degrees; whereas only 44% of the affiliated men had partners with advanced degrees. Forty-four percent of the female chemists versus 64% of the male chemists had children. Moreover, among all chemists with children, female chemists had fewer children than the males. On average female chemists had 0.8 children and the men had 1.3 children. In comparison, the average number of children per adult in Germany is currently 1.3. The nurturing of children less than three years is mainly provided by the partners of male chemists, whereas mothers who are chemists must organise a patchwork of care givers consisting of parental care, day care, and nannies, as well as, arranging for flexible working hours.

Discussion

The current situation faced by women and men in chemistry in Germany is profoundly different. While gender disparities have almost disappeared at the beginners' level when students are starting to study chemistry at German universities, men clearly outnumber women at the qualification level of *Diplom*. This means that no gender equality is reached as measured by an equal distribution on *all* qualification stages. A comparison between actual and projected percentages for a gender shows, however, that during the qualification phase any differences that exist are a result of dissimilarities in the choice of subject to study. The main schism between the genders occurs after the Ph.D.: the level of education required for higher positions in university and industry. On the basis of the relatively high percentage of doctorates in chemistry granted to women twenty years ago, fewer women hold these select positions than would be expected. A comparable situation is also observed in physics (*30*) and mathematics (*31*) in Germany, as well as, in chemistry in the US (*32, 33*).

The lower numbers of women in leading positions cannot be explained solely by the fact that women still bear the major responsibility for household and child care. Of course, it is very important to improve public child care especially in Germany where public child care and elementary schools are open mostly only in the mornings, thus leaving the children completely to their parents' responsibility in the afternoons. It is also important to change the public's perception of viewing motherhood as incompatible with demanding professional careers. The fact is that women *without* children have not attained positions as prestigious as those held by men. This makes clear that the "fact of potential motherhood conveys unfavorable effects onto the implicit achievement potential and further career development." (*34*) Another explanation for women not holding higher management positions is that women more readily than men defer their professional ambitions in favor of their partners who generally have demanding careers, thereby complicating the common professional and private life of the couple. The public discussion of the roles of such dual-career couples is only just beginning in Germany (*35-37*).

What should be done? Two evident areas of action can be distinguished:

- The efforts to increase college enrollment launched in the recent years should be continued while taking care that female and male high-school students are optimally addressed. This should help in guaranteeing that all students interested in chemistry enroll as undergraduates, irrespective of their gender.
- The subtle and gradually increasing disadvantages that women face in professional contexts must be addressed openly and gradually abolished.

Social Implications

What is equal opportunity? Without generalization, equal opportunity means that all social players have the possibility of fully developing their respective individual potentials independent of gender or whether they belong to specific social groups. In order for a person to develop her or his individual potential, the conditions must be such that optimal support is guaranteed. If one classifies society into groups such that each member belongs to only one group and that each group is given the opportunity to reach their potential (*e.g.*, the ability to "think chemistry"); then the social environment must allow the potential of a member to be developed independent of the member's social group. If this criterion is met, equal opportunity with respect to the development of the specific potential is given. However, as soon as the development of a given potential is supported for members of one social group but hindered in another, unequal opportunities arise.

Nobody will doubt that women today, just like men, have the potential to do scientific work and to effectively hold responsible positions. The data presented in this article shows unambiguously that women and men have an equal interest in chemistry as a science. Thus, if society is classified into social groups as "women" and "men", one must assume an equal distribution within those two groups with respect to their *potential* to work effectively as a chemist at universities or industry. However, clearly, there is no equal distribution of the genders in leading positions in either work sector. The question as to whether there is equal opportunity in chemistry can be thus answered with a clear: No!

There has been much discussion on the personal and social costs of having gender inequality. The source of the imbalance is not well understood nor are the remedies defined. The fallacy of a commonly used argument is demonstrated in a study conducted at BASF, Germany: "Often I hear in our company that women cannot go into the operative business because of their life and family planning. With having a family, they would change their plans for their lives and stay at home or work part time. A survey of internal data within the BASF shows, however, that highly-qualified women tend to shorten their entitled child-care leave and use part-time options only sparingly and with a higher number of working hours." (*38*)

Role expectations held by the public, however, certainly play a significant part. This was stated by Linda Austin in her book "What's holding you back?" where she writes: "While courage is surely an important trait for the achieving man, women must be even more psychologically brave than their male counterparts to succeed. After all, it is so clearly within the scope of the expected male behaviour to take independent, autonomous action. [...] For women, boldness puts her distinctly at odds with the role that society expects of her. " (*39*)

This quotation shows that there is a difference in the expected behaviour of women and men. The conclusion that implicit criteria used in judging people is the rational for the observed unequal opportunities between the genders also has been found by the Bund-Länder-Kommission: "Assuming that talents, qualification, and professional aims of men and women are not significantly different, the over-proportional share of men suggests that – among several possible factors – different achievement and qualification measures are used." (*23*)

The results of statistical surveys represent a "social average", resulting from the individual life prospects of the persons questioned. If the situation were to be changed - as evidenced by a change in the statistics – a sufficiently large proportion of all persons polled would have needed to change their personal life styles. As long as being female – however this may be connatated – is consciously or unconsciously perceived by a large part of the general population as being incompatible with success in chemistry, and being male is perceived as a necessary condition for success in chemistry, the current situation will change only slightly. However, Germany being a country with few natural resources needs adults of both genders who can combine the roles of being both innovators and parents. If any of the two roles is not adequately fulfilled on a massive scale, a social catastrophe may follow: if, on the one hand, there is a lack of highly qualified innovators, the necessary conditions for maintaining the living standards of a modern industrial society will deteriorate, resulting in receding quality of life. If, on the other hand, too few children are being born, resulting in a demographic collapse, there will be a deterioration of the current social order – a trend that is already observable in Germany. Thus, one must demand that both genders have equal opportunity to obtain highly qualified professional work and that there is a sharing of family responsibilities in a just way, *i.e.*, by making optimal use of the existing individual potentials of all people involved. This however, calls for a change in attitude by *all* social actors.

Latest investigations show that being engaged in family issues is not at all a hindrance for professional work. On the contrary, being involved in two different sectors has been found to lead to personal enrichment and a more flexible repertoire of action that can be used for creative and innovative organisation of professional work. The social scientist, Hildegard Macha, conducted a biographical study on successful female scientists in Germany, in which she writes: "By drawing energy and confirmation out of their roles in different life contexts and developing a high competence to deal with ambivalences and multiple requirements they can better cope with their daily routines in their family and profession, they show higher self-esteem and well-being. These results could also form the basis for new holistic life models for men who in the past have concentrated mainly on the professional sector – often at the cost of their health – and have neglected the private sector with its compensation potentials." (*40*)

The social challenge is to develop, communicate, and institutionalize an integrative approach that replaces the currently predominant "either-or" approach. The personal challenge for female chemists is to reflect acquired values, normative systems, and role expectations and to actively shape their lives. Quoting Linda Austin once again: "... achievement requires the ability to determine a unique sense of meaning, often radically different from what society suggests. [...] She must develop her own version of what it is to be female." (*41*) However, it would be unilateral to review and discuss only the situation of women, even if the focus of the current discussion is on women. As life changes for women, men will have to adjust. How the change of expectations, as well as life prospects, affects the self perception and career choices of men is a question gaining interest only recently. Changes in the working environment and in family life will certainly affect both genders. Thus, the AKCC strongly encourages that more men become involved in the discussion on gender equality.

References

1. http://www.gdch.de/strukturen/fg/akcc.htm
2. *Nachr. Chem.* **2000**, *48*, 1149.
3. http://www.vdi.de/fib
4. VDI brochure *20 Jahre Frauen im Ingenieurberuf*; VDI 2002.
5. http://www.dibev.de
6. http://www.physikerin.de
7. Zilles, D. *Phys. Bl.* **1999**, *55*, 9.
8. Osborn, M.; Rees, T. ; Bosch, M. ; Ebeling, H. ; Hermann, C. ; Hilden, J. ; McLaren, A. ; Palomba, R. ; Peltonen, L. ; Vela, C. ; Weis, D. ; Agnes Wold, Mason, J.; Wennerås, C. *Science policies in the European Union. Promoting excellence through mainstreaming gender equality*. Office for Official Publications of the European Communities: Luxembourg, **2000**, p 7.
9. http://www.gdch.de
10. http://www.destatis.de
11. http://www.vaa.de
12. Schmitz, K. *Nachr. Chem.* **2004**, *52*, 801-804.
13. http://www.uni-bielefeld.de/teutolab. *Nachr. Chem.* **2000**, *48*, 702.
14. Jennett, H.; Kohse-Höinghaus, K. *Nachr. Chem.* **2003**, *51*, 144ff.

15. http://www.bildungsinitiative-chemie.de
16. http://www.uni-frankfurt.de/didachem/gdchfb
17. http:// www.gdch.de/vas/fortbildung/lehrer.htm
18. *Chemie in unserer Zeit* **2000**, *34*, 182. Compare also the press release *Chemie fördert die die besten Abiturienten*; VCI: February 9, 2000.
19. http://www.vci.de/
20. Brochure *Schulpartnerschaft Chemie*; Fonds der Chemischen Industrie im Verband der Chemischen Industrie e.V.: May 2003.
21. http://www.ada-lovelace.de
22. Ebeling, H.; Dewandre, N.; Douka, M. *Nachr. Chem.* **2003,** *51*, 30-32. The quote has been translated by the author.
23. Bund-Länder-Kommission. *Frauen in der Wissenschaft – Entwicklung und Perspektiven auf dem Weg zur Chancengleichheit*; Bonn 2000, p 8. The quote has been translated by the author.
24. Nägele, B. *Von "Mädchen" und "Kollegen". Zum Geschlechterverhältnis am Fachbereich Chemie*; Talheimer Verlag: Mössingen-Talheim 1998.
25. *VAA-Nachrichten* September, **1995**, p 2.
26. *VAA-Nachrichten* September, **2001**, p 4.
27. Rübsamen-Waigmann, H.; Sohlberg, R.; Rees, T.; Berry, O.; Bismuth, P.; D'Antona, R.; de Brabander, E.; Haemers, G.; Holmes, J.; Jepsen, M.; Leclaire, J.; Mann, E.; Needham, R.; Neumann, J.; Nielsen, N. C.; Vela, C.; Winslow, D. *Women in Industrial Research. A wake up call for European Industry*; Office for Official Publications of the European Communities: Luxemburg, 2003.
28. Könekamp, B. *Nachr. Chem.* **2004,** *52*, 147-153.
29. Könekamp, B.; Krais, B.; Erlemann, M.; Kausch, C. *Physik Journal* **2002,** *1*, 22-27.
30. Bessenrodt-Weberpals, M. *Physik Journal* **2003**, *2*, 31-35.
31. Abele, A. E. In *Frauen und Männer in akademischen Professionen. Berufsverläufe und Berufserfolg*; Abele, A. E.; Hoff, E.-H.; Hohner, H.-U. Eds.; Asanger Verlag: Heidelberg Kröning, 2003, pp 97-112.
32. Kuck, V. *Chemical & Engineering News* **2001**, *79*, November 19, 71-73.
33. Heylin, M. *Chemical & Engineering News* **2004**, *82*, February 16, 68-72.
34. Lind, I. In *CEWS Newsletter Nr. 26*, **2004**. http://www.cews.uni-bonn.de/ . The quote has been translated by the author.
35. Abele, A. E. In *Frauen und Männer in akademischen Professionen. Berufsverläufe und Berufserfolg*; Abele, A. E.; Hoff, E.-H.; Hohner, H.-U. Eds; Asanger Verlag: Heidelberg Kröning, 2003, pp 157-182.
36. Brochure *Dual Career Couples. Karriere im Duett. Mehr Chancen für Forscherpaare*; Deutsche Forschungsgemeinschaft und Stifterverband für die Deutsche Wissenschaft: March 2004.
37. http://www.dfg.de/wissenschaftliche_karriere/focus/doppelkarriere_paare/

38. Brochure *Vielfalt und Chancengleichheit. Einblicke in die Praxis der BASF*; VAA: 2002, p 33. The quote has been translated by the author.
39. Austin, L. *What's holding you back?* Basic Books: New York, 2000, p. xviii.
40. Macha, H. und Forschungsgruppe. *Erfolgreiche Frauen. Wie sie wurden, was sie sind*; Campus Verlag: Frankfurt New York, 2000, p. 246. The quote has been translated by the author.
41. Austin, L. *What's holding you back?* Basic Books: New York, 2000, p. xix.

Chapter 8

Findings from the American Chemical Society Career Continuity Survey: Elucidating Gender Patterns in Training and Career Paths

Cecilia H. Marzabadi

Department of Chemistry and Biochemistry and Center for Women's Studies, Seton Hall University, 400 South Orange Avenue, South Orange, NJ 07079 (email: marzabce@shu.edu)

The purpose of this study was to examine the career paths of a group of graduates from prestigious chemistry programs in the United States. Surveys were mailed to male and female doctoral graduates from the top-ten chemistry programs, as defined by the National Research Council (NRC), for the years 1988-1992. Based upon the participants' survey responses, possible gender-based barriers to success in educational and training environments were elucidated. The further impact of these barriers to the early career paths of women, particularly to their selection of and placement on the faculties at Ph.D.-granting universities, also was examined. Best practices to encourage and promote women in these settings are proposed.

Introduction

Over the past several decades, the presence of female faculty members in the chemistry departments of four year colleges has increased steadily. For example, in the year 2000, women made up 11.8% of full professors, 25.8% of associate professors, and 34.2% of assistant professors in chemistry departments granting B.S. terminal degrees (*1*). This compares with 6.7%, 18.0% and 25.2%, respectively, one decade earlier. However, the numbers of women on the faculties at Ph.D.-granting institutions have increased much more slowly over the same period of time. In these schools women constituted only 7.9% of the full professors, 18.0% of the associate professors, and 25.2% of the assistant professors in the year 2000, up from 4.3%, 12.3% and 18.4% , respectively, in 1990 (*1*). Looking only at the top-fifty Ph.D-granting schools that spent the most on chemical research, as identified by the National Science Foundation (NSF) , the female faculty constituted only 12% of the total faculty in the period 2002-2003; up slightly from 11% in the previous year. This slow increase in female faculty has occurred in spite of the fact that the percentage of women receiving Ph.D.s in chemistry has increased to more than 30% of all graduates (*2*). In fact, an analysis of the faculty composition in various STEM (science, technology, engineering and mathematics) fields has shown that among the top-ten departments of chemistry, physics, and engineering, only chemistry departments are hiring disproportionately below the available applicant pool of women (*3*).

The reasons for this low representation of women have not been fully investigated. It is not clear whether gender discrimination in the workplace, the choices of female scientists, or a combination of these two factors is responsible for the low placement of women in academic positions in chemistry. It has been suggested that women have not been applying for these positions in the proportions in which they are graduating from the top schools (*4*). A hostile graduate school environment with poor student mentoring and few, if any, same sex role models may be affecting their decisions.

Issues of isolation, lack of direction and lack of professional contacts plague women throughout their graduate careers (*5*).Their scientific contributions are frequently devalued (e.g. in group meetings) and they are often excluded from professional events (e.g. conferences). Thus, women often find it difficult to be taken seriously. Because they are made to feel uncomfortable, women hold back their scientific opinions, which leads to further isolation. Women may therefore be reluctant to ask for help, because then they may be labeled as dependant. All of these factors contribute to a lower sense of self-confidence and impede the creative process in women. This may discourage women from seeking top academic positions or adversely affect the subsequent interview process for these candidates. Taken together, these factors ultimately may lead to the attrition of women from the field.

A lack of relevant role models for women beginning academic careers in chemistry also may be deterring them from pursuing these occupations. Even in departments where the numbers of female faculty members have begun to increase, frequently a subdivision into two distinct groups can be seen. Some women typically share the values and work styles of the older men. In contrast, some women (and men) struggle to create an alternate scientific lifestyle, balancing work and other life issues (*6*).

The desire for a balanced lifestyle coupled with the required long work hours, the pressures of the tenure process, and the toll that these factors take on personal and family life, all may be contributing reasons for the low application rates by women for positions at Ph.D.-granting institutions. For women considering starting families, often the timing of the initial academic appointment runs parallel to their "biological clocks." Because the academic culture in science departments often views a commitment to work and a commitment to family as two opposing ideals, many women may be opting out of this environment for personal reasons.

Women have made more significant inroads into academe at less prestigious institutions. This may be happening as they perceive working at a four-year college (or other, less research-intense institution) as more compatible with having a family. Some women may be choosing industrial positions because of the greater access to child care facilities and the perceived flexibility in work hours.

Often female chemists marry other scientists or professionals (*7*). At the completion of their degrees, this results in a "two-body" or "dual-career" problem that also may be contributing to lower numbers of women in academe. The ability to locate two rewarding jobs within the same geographical region may present a challenge that leads to career or relationship compromises for a dual-career couple. Frequently women narrow the scope of their geographical job searches to accommodate their partners (*8*). In academic fields, in which there are limited numbers of positions and in which one must be geographically-flexible when searching for jobs, this can be a serious impediment.

Yet another factor to consider when looking at the low numbers of women on faculties at Ph.D.-granting institutions is whether female graduate students and post-doctoral fellows are receiving the same quality and quantity of training and mentoring as are their male counterparts. Poor relationships with doctoral and post-doctoral advisors can lead to feelings of incompetence and isolation. Also, the perception that the women are less committed than are men to the science (as based upon their choices for work/life balance) may lead to less active advocacy by mentors for their female graduate students and post-docs than for their male graduate students and post-docs (*5*).

In a previous chapter in this book, men's and women's perceptions of the mentoring they received as undergraduate and graduate students, post-doctoral fellows, and in their initial employment are discussed (*9*). The results presented suggest that women, as compared to men, indeed feel less included and advised, report being less likely to obtain top employment positions, and express lower satisfaction with their training and careers. In this chapter, we further compare the educational experiences of men and women to elucidate the factors that may be responsible for the under-employment of women in tenure track faculty positions at the top chemistry departments. The analysis of survey data from graduates from these top departments is important because they are the preferred applicant pool for faculty hires. For example, we have found that at the top-ten schools, 77% of the younger faculty (those that received their Ph.D.s after 1979) obtained their doctoral training at a top-ten school (*10*).

After re-examining aspects of their undergraduate and graduate education, we looked at the impact of additional post-doctoral training on the career paths of this cohort of men and women. We also looked at the numbers and types of applications for first positions that they made. The goal of this inquiry was to see if the women made applications to R1 (Research 1 – Carnegie classification) institutions and if so, whether they sought such positions in the same proportion as their male classmates. We also wanted to determine if the women preferentially sought positions at four-year colleges and other institutions and what the reasons for these career choices were. We looked at the number of interviews received by both the men and the women for all types of academic positions. Because all of these individuals should in principle have the same level of quality training, they should be equally qualified for a tenure track position at an R1 institution and should be receiving comparable numbers of interviews. Finally, we looked at the number and types of academic offers garnered by all of these individuals to begin to address whether the interview process has failed women in this cohort.

Methods

Participants

Participants in this study included a self-selected group from all women and men who had received a Ph.D. between the years 1988 and 1992 from one of the top-ten ranked chemistry programs, as identified by the National Research Council (NRC). These top-ten universities were (in ranked order): University of California, Berkeley; California Institute of Technology; Harvard University; Stanford University; Massachusetts Institute of Technology; Cornell University; Columbia University; University of Illinois-Urbana-Champaign, University of Wisconsin-Madison, and University of Chicago. Participants also included individuals from an eleventh program, graduates in the same five year period from a large, public, Midwestern university (Purdue University) which is not ranked as a top-ten school, but is ranked as the eleventh "supplier" of doctoral graduates who seek and attain academic employment at a top 50 school (*10*). All universities contacted for the study agreed to participate.

The particular five-year range utilized in this study was chosen because graduates in this group would have had ample time to complete post-doctoral training, begin their careers, and become familiar with the demands, expectations, and obligations associated with their first employment positions. We also expected that because not a lot of time had elapsed since their doctoral and post-doctoral training, that these individuals would be in a position to comment retrospectively on the reasons they used to make their early career choices.

Surveys were mailed to 1950 doctoral graduates in the spring of 2002. Of this number, 283 were returned unopened by the post office. Of the remaining 1667 questionnaires, 454 were completed and returned for a 27.3% response rate (*11*).

Materials

A letter from the president of the American Chemical Society (ACS) was sent to all departmental chairs/deans of participating institutions. The letter outlined the current study as an investigation of the training and career development of chemists, indicating the participating university's anonymity and the receipt of Institutional Review Board approval for the study. Moreover, the letter stated that the study had the full support of the ACS and it encouraged participation by all institutions to ensure completeness of data. Institutions were invited to mail a questionnaire and letter of solicitation directly to each of the graduates from the selected time frame (1988-1992), or to release the mailing addresses of their alumni to the investigators of this study for subsequent mailing. Similar letters from the ACS president were addressed to all potential participants in the study, describing the study as above, indicating that the participant's name was received from her/his institution, avowing ACS's support of the study, and encouraging participation.

A four-page questionnaire that we developed was used to survey the training and career choices that individuals made, as well as the reasons participants gave for making these choices. Specifically, we asked participants to evaluate their undergraduate, graduate, post-doctoral, and career experiences, and to compare their experiences to others in their peer group.

Most questionnaire items involved closed-ended choices; often, respondents were asked to rank their experiences relative to others in their peer group.

Frequently, a scale of 1 to 7 was used with 1 corresponding to "little," "minor," or "worse than," 4 corresponding to "neutral," "same as," or "neither worse than nor better than" and 7 "a lot better" or "major." There also were several open-ended questions that invited participants to provide numerical data and additional comments where appropriate. In this particular study, we were interested in survey responses related to the career choices made by individuals and the reasons for these choices.

After providing information about their undergraduate and graduate training experiences, survey respondents were asked questions about their post-doctoral fellowships and first employment positions. Information on the number of post-doctoral fellowships and the length of time they held these post-doctoral positions was obtained. They were then queried about their first positions. Individuals were asked how many positions they sought in different types of employment (e.g. Ph.D. granting institution versus non-Ph.D. granting institution, business/industry versus government) and whether the academic positions for which they applied were tenure track. They were asked to disclose the numbers of interviews they received from these applications and the number of employment opportunities that ensued from this process. Finally, participants were asked to comment on the reasons they used to make their career choices and to comment in hindsight on the correctness of their choices.

The Statistical Package for the Social Sciences (SPSS) software was used to analyze the data obtained from this questionnaire. For most analyses, we used *t*-tests for equal interval dependent variables, and chi square tests for independence for categorical dependent variables.

Results

Profile of Participants

The respondents to this survey consisted of 315 men and 135 women (4 respondents did not identify their gender). The gender breakdown for this group was 70% male and 30% female. These percentages were very close to the gender composition for the chemistry doctoral pool for the period 1988-1992 which had an average percentage of 28.2% female (*12*). The majority of the respondents to this survey were US citizens; 89% of the women and 86% of the men.

In our sample, 361 respondents reported that they were married (80.4%). About three-quarters (75.4%) of the women currently were married, versus. 82.5% of the men. Fifty-seven percent of the women and 69.8% of the men reported that they had one or more child.

Undergraduate and GraduateTraining

In order to accurately describe the impact of training on the career paths of this select group of individuals, it was essential to look at the training environments of these individuals from early on. The impact of undergraduate and graduate training have been reported previously (*13*) and will be summarized below. First, we examined the type of undergraduate institution that participants in our study attended. We found that women were more likely to have attended a four-year college than were men. Practically all of the graduates reported having had research experience as undergraduates.

When deciding which graduate school to attend, many of the respondents reported having had help making this decision. However, we found that men were more likely than were women to report that an undergraduate professor had helped them choose a graduate school. Women reported more frequently than did the men that they had "no help" in choosing a graduate school. When we asked participants to provide the gender of the person who helped them to choose a graduate school, women were more likely than were men to report having received help from a female faculty member.

All of the individuals surveyed subsequently obtained doctoral degrees from prestigious R1 institutions. Therefore, it would seem likely that they had received on average the same quantity and quality of training. Similar criteria were used by both the male and female respondents in choosing a dissertation advisor in graduate school. These included: the advisor's scientific reputation, publication record, funding record and reputation as an advisor. Once paired with a mentor, women and men did not report significant differences in the access they had to their mentor or in the amount of basic help that they received from their mentor in learning laboratory techniques and understanding theses requirements. So in quantitative terms, the students of both genders appeared to be on equal footing. However, when asked about the receipt of help in more specific areas such as: learning how to conduct independent research, evaluating data, achieving a teaching/research balance, developing career goals and finding a job, men rated higher than did women the help that their dissertation advisor provided in all of these categories. Men also gave higher ratings to their interactions with their dissertation professors than did women and gave higher ratings than did women to the overall interest that their dissertation advisor showed in them.

Men's and women's differing responses to questions about their perceptions of their training experience are worth noting. Although the male and female respondents may have been receiving similar help in learning the requisite skills to continue in their studies and to become chemists, they may not have been receiving the same quality of mentoring from their advisors.

The continued availability of mentors at the graduate and post-doctoral levels also is likely to be important. Because women reported seeking more help from female faculty members at the undergraduate level, it seems likely that this preference would continue at the graduate level. In our study, this trend was indeed observed. Female graduate students reported that they considered the genders of their advisors as a more important criterion for their selection than did men. Recently, Schlegel (*14*) offered an explanation for this preference; female graduate students were more likely than were male graduate students to view personal characteristics (i.e., lifestyle, values) of a same-gender mentor as integral to their own training and career development. The paucity of female faculty members in the graduate departments we surveyed (in 1997, there were a total of 20 women on the faculty at the top ten chemistry departments) likely provided few choices of same-sex mentors for the women in this study. Furthermore, we are unable to access whether the female faculty in these departments possessed the attributes that were desired by the female students in terms of lifestyle and personal aspirations. Quite possibly this made the choices of mentors even more limited for the surveyed group of female graduates.

The absence of senior women available to mentor female graduate students, post-docs, and junior faculty members may be leaving many women without a necessary support network (*15*).This dearth of female faculty may in part serve to explain the higher percentage of women than men in our study who wished that they had used different criteria to select their advisors. Similarly, a higher percentage of women than men stated that they actually did switch advisors during graduate school.

Post-Doctoral Training

The majority of the individuals who responded to the questionnaire continued their formal training as post-doctoral fellows. One hundred eighty five men (59.5%) and 85 women (63%) responded that they had taken at least one post-doctoral position following the completion of their Ph.D.s. The primary reasons for seeking a post-doc for both men and women were: 1) to develop skills, and 2) as a prerequisite for a job. When asked what criteria were given the most importance when choosing a post-doctoral position, men rated higher than did women the importance of the scientific reputation of the advisor (64.7% vs. 45.2%, respectively). Women cited geographic constraints as one of their primary motivators for choosing their post-doctoral positions (32.9% of the women vs. 15.5% of the men). In this cohort, men spent on average 1.7 years as a post-doctoral fellow and women spent on average 1.6 years in these positions.

These data are significant in light of some of our previous findings which revealed that a large percentage of the current faculty at the top-fifty NRC-ranked chemistry departments had held post-doctoral fellowships prior to their academic appointments (*10*). For faculty who had received their Ph.D.s after 1979, 90% of the men and 94% of the women had additional post-doctoral training. In addition, 53% of these younger faculties had their initial post-doctoral fellowship at a top-ten school.

Furthermore, women appear to be using different criteria to select than post-docs than are men. A larger percentage of the women used geographic considerations as one of their primary reasons for choosing their post-doctoral fellowships. This may be leading to less-prestigious post-doctoral positions for the women than for the men and may be differentially affecting their future career outcomes.

In order to look at the issue of dissertation advisor advocacy for the male and female graduate students, we queried the respondents about the helpfulness of their dissertation advisors in ensuring that they obtained the post-doctoral fellowship that they most desired. Again, men perceived that their advisors were more helpful than did women (1.92 vs. 2.23, with 1 being most helpful and 5 being unhelpful). However, both men and women reported receiving approximately the same number of offers for post-doctoral fellowships (~2) and a large percentage of both genders stated that they got the position they wanted (88.7% of the men, 91.6% of the women).

Previously we reported other data from this survey (*13*) that showed that once matched with a post-doctoral advisor, there were several gender differences in the amount of interest that participants' reported their advisors showed in a range of areas related to their training. For instance, men reported more advisor interest than did women in their research ideas and findings. Men also gave higher ratings than did women to several other aspects of their post-doctoral experiences, including their interactions with their advisors and the availability of publication opportunities, as compared to other post-doctoral peers. These findings mirror those reported for the graduate school experience of men and women in this cohort.

First Employment Position

Of the 315 men surveyed, 306 of them (97.1%) were currently employed; of the women 131 (97.0%) were employed. Twenty seven (8.6%) of the men and 17 (12.8%) of the women reported having been unemployed at some point in time since the completion of their doctoral studies. Of those who had been unemployed, women were more likely than were men to say it was voluntary (64.7% of the unemployed women versus 32.0% of the unemployed men). Reasons cited for voluntary unemployment included professional education for a different career path and family needs.

When asked whether they later sought full-time, part-time or no employment following their unemployment, no men reported looking for part-time employment, whereas 18.8% of the women that responded to this question said they had sought part-time work. This difference in responses across the genders may be explained by continuing family-related issues for the female respondents and the necessity to balance work and personal responsibilities. More than 91% of the men (91.3%) said they sought a full-time job, and 8.7% did not seek another position in chemistry. Quite differently, from the 87.2% of the women, 81% sought full-time jobs and 6.2% sought no job. Among respondents, women reported working part-time for much longer periods of time than did men (average of 19.0 months for women, versus 2.7 months for men).

These and other observed gender patterns that we discovered perhaps reflect the larger role that women play in the family-care process. For instance, when questioned about their first employment positions, only 1.0% of the men, but 5.6% of the women, reported taking part-time work after the completion of their training. Women also were more likely to have accepted temporary positions (23.4%) versus the men (13.5%). Reasons stated for accepting these part-time and temporary jobs included: lack of employment options, geographical constraints, pursuing professional interests other than chemistry, hopes of obtaining subsequent full-time/permanent positions, lifestyle issues, and family necessity.

The 315 male survey participants and 135 female survey participants also were asked how many job applications they completed, how many on-site interviews they received, and how many offers of employment they garnered for a range of academic, industrial, and government positions. The number of respondents of each gender in each of the aforementioned categories are given in Table I

Regarding academic jobs, more men applied for, interviewed for, and received job offers at colleges and universities at the completion of their formal training. About twenty six percent (25.9%) of the women that took this survey reported having submitted applications for tenure track positions at Ph.D.-granting institutions. This percentage of women is lower than the average percentage of women in the Ph.D. pool for this same period of time (28.2%). Men and women, on average, applied for approximately seven tenure track positions at Ph.D.-granting institutions and they interviewed at only one of these seven schools. Approximately seventy-two percent (72.1%) of the male applicants reported having received interviews following their applications and 76.2% of those that interviewed (55% of the male applicant pool) reported having received one or more employment offer for a tenure track positions at a Ph.D-granting institution.

Table I. Number of Respondents Who Applied, Interviewed, and Received Offers in Different First Employment Choices

Positions Sought	Gender	Applied	Interviewed	Offers
Ph.D. tenure track	Male	111	80	61
	Female	35	26	24
Ph.D. non-tenure track	Male	24	15	16
	Female	19	16	11
Non-Ph.D., tenure track	Male	63	45	36
	Female	31	24	23
Non-Ph.D., non-tenure track	Male	24	19	13
	Female	15	17	16
Industry , small company	Male	61	45	34
	Female	16	12	8
Industry, medium company	Male	87	59	40
	Female	16	11	8
Industry, large company	Male	194	172	151
	Female	58	58	50
Government	Male	66	45	35
	Female	20	15	15
Other	Male	9	6	8
	Female	14	10	10

A high percentage of the female applicants, likewise, were granted interviews (74.3%), however, more of the women (92.3%) reported having successful interviews for these positions. Overall, 69% of the female applicants received offers for tenure track positions at these institutions. For non-tenure track positions at Ph.D-granting institutions, the opposite trend was observed. Despite the almost two-fold higher preference by women for these positions (14.1% of the women in the survey pool applied for these jobs versus 7.6% of the men), a higher percentage of the male applicants reported that they had actually received offers of employment , compared to the women that applied to these same institutions (66.7% for the men versus 57.9% for the women).

From all respondents to this survey, a smaller percentage of the men than the women (27.6% versus 34.1%) said that they had applied to non-Ph.D. granting institutions for both tenure track and non-tenure track positions. Among those who applied, men reportedly submitted more applications than did the women (for non-Ph.D tenure track, men submitted on average 3.14 applications versus

1.90 average applications for the women), but women reported receiving more employment offers at these schools than did men (at these same schools, men received on average 0.18 offers versus 0.40 average offers for the women). Only little more than half of the male applicants (56.3%) versus more than three quarters (84.8%) of the female applicants reported being successful at securing an offer for a tenure-track position at a non-Ph.D granting institution.

When respondents were asked whether their first employer was academic or non-academic, 32% (100) of the men and 47% (62) of the women replied that they took their first position in academe. The type of institution they gave as their initial employment and the percentage of each gender reporting being employed in those institutions is given in Table II. These numbers and percentages reflect the total for both tenure track and non-tenure track positions.

Table II. Type of First Academic Position

Institution Type		**Male**	**Female**	**Total**
Doctoral Extensive	Frequency Percent	49 49.0%	23 37.1%	72 44.4%
Doctoral Intensive	Frequency Percent	11 11.0%	7 11.3%	18 11.1%
Master's	Frequency Percent	10 10.0%	3 4.8%	13 8.0%
BA	Frequency Percent	22 22.0%	24 38.7%	46 28.4%
BA or Associates	Frequency Percent	0 0.0%	1 1.6%	1 0.6%
Medical School	Frequency Percent	3 3.0%	1 1.6%	4 2.5%
Other	Frequency Percent	5 5.0%	3 4.8%	8 4.9%
Total	Frequency Percent	100 100.0%	62 100.0%	162 100.0%

Respondents were asked to classify the type of instiution at which they took their first employment according to the highest degree awarded by the institution or as a medical school or other. Within the doctoral institutions, they were asked to further clarify whether it was a doctoral extensive or doctoral intensive school as defined by the Carnegie classification system. Doctoral extensive institutions are defined as those that award 50 or more doctoral degrees per year across at least 15 disciplines. The doctoral intetensive schools are those that award at least ten doctoral degrees per year across three or more disciplines, or at least 20 doctoral degrees per year overall.

An initial examination of this data suggests no statistically significant difference in the types of first academic positions held by men and women. However, when these data are further broken down into those who were hired into predominately graduate (M.S. or Ph.D.-granting) institutions in contrast to those who were hired into predominately undergraduate (BA or Associates) institutions, a statistically significant effect across genders is observed. Excluding those who obtained employment at medical schools or at other institutions, 70.0% of the men who took an academic position reported receiving their first employment at a predominately graduate institution, whereas only 53.2% of the women reported doing so.

We further examined this data looking only at those individuals whose first employment was at tenure track positions at Ph.D.-granting institutions. From the data presented in Table II, 61 men and 24 women reported receiving employment offers at these schools. The breakdown of the first employment positions of these individuals is presented in Table III.

Table III. First Employment in Tenure Track Positions at Ph.D–granting Institutions

Type of Institution	Gender	First Employment Choice
Doctoral Extensive	Male Female	37 16
Doctoral Intensive	Male Female	8 5

Of the 61 men that were offered a tenure track position at a Ph.D-granting institution, 45 (73.8%) accepted such positions. For the women, 24 garnered offers, but only 21 (87.5%) accepted a position at either a doctoral extensive or doctoral intensive school. A higher percentage of the women (23.8%), than the men (17.8%), accepted employment offers at less-prestigious, doctoral intensive schools.

Several distinct phenomena appear to be occurring. Women are receiving a disproportionately higher percentage of job offers than are men, at non-Ph.D granting schools. Also, although women are having greater success at receiving offers for tenure track positions from Ph.D-granting institutions than are men, they are not applying for these positions in the same percentages in which they are receiving their doctoral degrees. Furthermore, women that are receiving tenure track offers are accepting more positions at lower-ranked Ph.D-granting schools.

A lack of "qualified" female applicants for tenure track positions is a common problem also often cited by hiring officials at top R1 institutions. The low number of women in our survey applicant pool supports this premise. In light of some of our prior data that showed the need for a post-doctoral fellowship in order to be hired at one of the top institutions (*10*), we believed it was important to look at the percentage of women who actually pursued this additional training before entering the workforce. In our survey, only 63% of the women reported having done additional post-doctoral work. This translates into a maximum of only 85 (63% of 135) women who would be considered qualified enough to interview for a position at an R1 institution; this is compared to 186 men (59.5% of 315). Furthermore, from our survey instrument we have no way to ascertain the type of institution from which the men and women received their post-doctoral training (e.g. from a top-ten school). Because the women report having had less geographical flexibility than did the men when making the choice of their post-doctoral position, it is likely that they applied and/or accepted positions from less-prestigious institutions than did the men. Because the NRC-50 schools tend to hire faculty that held post-doctoral fellowships from the top-five ranked schools (*10*), an individual who pursues a post-doc elsewhere appears to be a less "qualified" applicant for a faculty position at these institutions. This perception by the women (and men) that they are not "qualified" enough may be discouraging some individuals from submitting applications at these institutions. Responses to some of the open-ended questions in our survey supported the awareness of this perception by individuals in this cohort. These will be described below.

Considering other aspects of "qualification", it often is posited that the female applicants' training may be too specialized and may be an inappropriate "fit" for the positions available in Ph.D-granting departments. For those women in our cohort who applied for these positions and were unsuccessful in their endeavors this may be one possible rationale. However, recent data from the *Survey of Earned Doctorates* contradict this hypothesis and show that other than in the sub-discipline of organic chemistry, doctorates in physical, analytical, and inorganic chemistry are being obtained by women in equal or greater percentages than those by men (*16*).

We also prompted open-ended comments from these individuals about the reasons why they chose their initial career paths. Respondents to our survey were asked to give reasons why they did not pursue a tenure track appointment at a Ph.D.-granting institution. We received 51 responses to this question; 29 from women and 22 from men. From these responses, a large number of both women

(11) and men (9) stated that they did not choose this career path because of the pressure and lifestyle expectations associated with it. Specifically, one woman and one man responded that they did not choose jobs at these institutions because of the unfavorable view of the "tenure track" that they had observed while in graduate school. Another major response from this group was that they wanted to focus more on the teaching aspects of academe and felt they were more suited for a position at a two- or four-year college (10 women and 4 men). Other individuals did not feel they were qualified enough for these positions because they had not done a post-doc or did not have enough quality publications (2 women and 2 men). Three women and two men were dissuaded from making applications for tenure track positions at Ph.D.-granting institutions because of the lack of positions available to them. Other respondents simply had different professional plans or financial expectations that were incongruent with an academic lifestyle.

So, from the final, small pool of women (and men) who are being offered positions at Ph.D-granting institutions and are not accepting them, it is possible that they are opting out in favor of a position at a four-year college or elsewhere. They may be making these choices because of the potential conflicts they anticipate between work and life issues.

Finally, both genders were queried about how they found their first positions. Specifically, we were interested in knowing if they received help from either their dissertation advisors or post-doctoral advisors in this process. In both cases, a higher percentage of men than women said that these individuals helped them find their first positions; 9.2% of the men and 3.7% of the women received help from their dissertation advisors and 9.2% of the men versus 6.7% of the women received help from their post-doc advisors. In short, in addition to factors of uncertainty with the tenure process and their abilities, the desire to teach and lifestyle issues, there appears to be less support for women from their advisors to pursue these positions at R1 institutions.

Summary / Possible Best Practices

The focus of this paper has been on the low numbers of women on chemistry faculties at Ph.D. granting institutions in the U.S.. Results from a survey we administered to doctoral graduates (1988-1992) from 11 schools were summarized. Specifically, we examined training at the undergraduate, graduate, and post-doctoral levels and the impact of this training on future career outcomes, particularly in regards to careers in academe.

At all educational levels, women reported feeling more isolated and less satisfied with their training experiences. Women sought out same-sex mentors and role models at various stages in their training and frequently could not find these individuals. Often they reported that they did not want to emulate the lifestyles of the individuals with whom they had interacted in graduate school and as a post-doc.

Upon leaving graduate school, less than two-thirds of the women pursued a post-doctoral position, a necessary requirement for a faculty position at most institutions. Whereas men used the academic reputation as their primary reason for choosing their post-doc positions, women reported geographical considerations as a primary criterion when making their choices. The women reported feeling less supported by their advisor in graduate school and having received less help in securing post-doctoral positions. At the conclusion of their post-docs, women again felt that they had received less help and support from their post-doctoral advisors.

When looking for first positions, 5.6% of the women reported taking part-time positions and 23.4% took temporary jobs. Almost thirteen percent (12.8%) of the women reported having been unemployed at some time following their educational training. Only 35.6% of the women surveyed reported having sought a tenure track position. Although almost two-thirds of the women reported pursuing requisite post-doctoral studies, only about a quarter of the women in this pool applied for a tenure-track position at a Ph.D-granting institution. Major reasons for not applying for positions at R1 institutions included lifestyle issues and a preference for teaching versus research. On a positive note, it appears as if universities are attempting to hire a higher percentage of female applicants relative to the male applicants into the available positions in their chemistry departments.

The results obtained from our survey shine light on several problems that exist in the training of chemists for positions at academic research institutions. Although the discussions are particularly addressed toward attracting and retaining women in academic chemistry, in many cases, the comments are readily applicable to male students and post-docs pursuing these professions. First, women reported feeling less mentored than did men at all levels of their training. An increased presence of female mentors at all levels would be instrumental in retaining female trainees in the career pipeline for future positions. In addition, these female role models should possess qualities that reflect a balanced lifestyle (lifestyle versus work). Senior female faculty and all male faculty need to do more to mentor and advocate for their female (and male) students. This includes encouraging the women in their research groups to present their research at conferences, to do post-doctoral fellowships at top research institutions, and to pursue positions at research universities. Also, efforts should be made to inform and to advocate for those women interested in academic positions. The numbers (and percentages) of female applicants are still too low to make a difference in spite of increased efforts to bring them into the pipeline!

In order to realistically affect the number of women on their faculties, universities must also take measures. These might include making even more offers to qualified female candidates. This may have to be done even more disproportionately to the number of offers to men in order to see any real effect in the compositions of their departments. These top universities may have to extend their search efforts to individuals who have carried out post-doctoral training at institutions other than the most elite. Considerations for women desiring to start families either prior to or during their initial academic appointments must be made. Better family leave policies might be implemented, as well as making available funding for additional faculty lines or for other career assistance for spouses in order to attract good female candidates. Once hired into a department, additional and better formal and informal mentoring of female (and male) faculty may need to be implemented to increase their retention in the discipline.

References

1. *Chem Census 2000;* American Chemical Society, Committee on Economic and Professional Affairs: Washington, D.C., 2001; p 19.
2. Long, J. R. *Chemical & Engineering News.* **2002**, *80*, 110-111.
3. Kuck, V. J. *Chemical & Engineering News.* **2001**, *79,* 71-73.
4. Brennan, M. B. *Chemical & Engineering News.* **1996**, *74*, 7-15.
5. Etzkowitz, H.; Kemelgor, C.; Uzzi, B. *Athena Unbound, The Advancement of Women in Science and Technology*; Cambridge University Press: Cambridge, U.K., 2000, Chapter 6, 83-103.
6. Etzkowitz, H.; et. al. Reference 5, Chapter 11, 147-148.
7.. *Chem Census 2000;* American Chemical Society, Committee on Economic and Professional Affairs: Washington, D.C., 2001; p 10.
8. Preston, A. E. *Leaving Science. Occupational Exit from Scientific Careers;* Russell Sage Foundation: NewYork, N.Y., 2004; 73-81.
9. Nolan, S. *Dissolving Disparity, Catalyzing Change: Are Women Achieving Equity in Chemistry*; Chapter 4.
10. Kuck, V.J., Marzabadi, C. H.; Nolan, S. A.; Buckner, J. P. *J. Chem. Ed.* **2004**, *81*, 356-363.
11. Analysis revealed that our response rate likely was affected by the nonparticipation of one school. If adjustments are made for this school, the overall rate is 32.1%.
12. NSF-NIH Survey of Graduate Students & Post-doctorates in S&E.
13. Nolan, S. A.; Buckner, J. B.; Marzabadi, C.H.; Kuck, V. J. *Manuscript submitted to Sex Roles.*

14. Schlegel, M. *American Psychological Association Monitor.* **2000**, *31,* 33-36.
15. Riordan, C. A., Manning, L. M., Daniel, A.M., Murray, S. L., Thompson, P. B., & Cummins, E. *Journal of Women and Minorities in Science and Engineering.* **2000**, *5*, 29-52
16. WEBCASPAR, National Science Foundation's Survey of Earned Doctorates/Doctorate Records.

Chapter 9

Women in Academe: An Examination of the Education and Hiring of Chemists

Valerie J. Kuck

Center for Women's Studies and Department of Chemistry and Biochemistry, Seton Hall University, 400 South Orange Avenue, South Orange, NJ 07079–2694

An overview of the education and hiring of chemists is presented with an emphasis on the gender composition at each stage of education. The different career patterns of men and women are explored and possible reasons for the under-representation of women on the chemistry faculties at PhD-granting institutions are addressed.

Introduction

In 1969 the Title VII Civil Rights Act of 1964, which outlawed sex discrimination in employment, was amended by Executive Orders 11246 and 11375 to include higher education. Regulations for implementation of the amended bill were issued in 1972. While those ground-breaking events were occurring, several Ivy League universities started admitting women to their undergraduate and graduate programs. However, this was not the case for all of the elite universities. For instance it took Caltech (California Institute of Technology) almost another decade before it started accepting female applicants *(1)*. The decision of those prestigious institutions to admit women into graduate school was significant, as it gave women for the first time the opportunity to obtain the prerequisite credentials for consideration for faculty positions at the nation's leading research institutions.

In this paper, we examine the progress women have made in obtaining a greater portion of the baccalaureate and doctorate degrees granted in chemistry. We determine if women are holding post-doctoral appointments in proportion to their representation in the doctoral pool. Finally, we look at how well women have done in attaining tenured and tenure-track faculty positions at PhD-granting, MS-granting and 4-year institutions and compare those findings with the gender distribution of the individuals in non-tenure track positions. We conclude by identifying the doctoral institutions that trained the greatest number of female faculty members at the top 50 National Research Council (NRC-50) ranked chemistry departments.

Academic Training

Baccalaureate Degree

During the past five decades, the total number of bachelor degrees granted in chemistry has varied, reaching a high in the 1970s and then irregularly trailing off in subsequent decades (Table I). Over those years, the number of degrees earned by men has decreased whereas that for women increased. Those conflicting trends have accentuated the growth in the percent of baccalaureate degrees in chemistry earned by women. According to the American Chemical Society (ACS), in 2002 women received fifty percent of the bachelor degrees granted *(2)*. Another group growing in number is the foreign-born who according the National Science Foundation's *Science and Engineering Indicators-2004* received 14.9% of the bachelor degrees granted in chemistry in 1999.

After graduating from college, the usual pattern until the mid-1990s was that about 1/3 of the chemistry graduates went on to graduate school in chemistry, 1/3 entered other graduate programs, predominantly in medicine and dentistry, and 1/3 got jobs in industry or teaching *(3)*. In 1996 there was a significant shift in the after-graduation plans of many students. The proportion of students having no graduate plans increased substantially, while those going on to graduate school in chemistry decreased. In 1998, the ACS found that only 20% of the bachelor graduates planned on entering graduate school in chemistry and 52% had no graduate school plans. By 2001 there was a slight upswing in the percent continuing their studies in chemistry, as 24% of the baccalaureates indicated that they were going to graduate school in chemistry, 29% to other graduate studies and 47% had no graduate school plans.

Table I. Baccalaureate Degrees in Chemistry, 1950-2001

	Degrees Awarded			
Year	*Total*	*Men*	*Women*	*% to Women*
1950	8696	7392	1304	15.0
1960	7604	6096	1508	19.8
1970	11617	9501	2116	18.2
1980	11446	8169	3277	28.6
1990	8289	4965	3324	40.1
2000	10390	5483	4907	47.2
2001	9822	5047	4775	48.6

Source: NSF, /Division of Science Resources Statistics; data from Department of Education/National Center for Education Statistics.

Graduate School

Since 1980 the number of students entering graduate school in chemistry has varied (Table II). Similar to the pattern observed with the baccalaureate degree recipients, the number of men has decreased slowly; whereas, that for women has increased significantly, going from 24.8% in 1980 to 43% of the first-time, full-time graduate students in 2002. It should be noted that the percentage of women in the entering graduate school classes has continued to lag behind their fraction in the baccalaureate pool.

Table II. Gender Distribution of the First-time, Full-time Chemistry Graduate Student Pool, 1980-2002

Year	*Total*	*Men*	*Women*	*% Women*
1980	3,394	2,551	843	24.8
1985	3,673	2,641	1,032	28.1
1990	3,655	2,398	1,257	34.4
1995	3,564	2,294	1,270	35.6
2000	3,374	2,031	1,343	39.8
2002	3,658	2,084	1,574	43.0

Source: NSF-NIH Survey of Graduate Students & Post-doctorates in S&E.

Not only has the gender distribution but also the citizenship of the graduate students been changing over the years. Unfortunately, the National Science Foundation (NSF) only started tracking the citizenship of the entering graduate students in 2000. To get a rough estimate of the number of U.S. citizens and permanent residents entering graduate school for a given year, one can use the citizenship distribution of the doctorate pool six years later. Assuming that reception of doctorate is independent of citizenship, foreign students should have comprised around 25% (an extrapolated value) of the entering class in 1980 (Table III). In succeeding years the fraction of foreign students in the entering classes should have increased based on their fraction of the doctoral degrees.

Table III. Fraction of PhD Degrees in Chemistry Awarded to Foreign Students

	Number of PhD Degrees Granted In Chemistry		
Year	*Total*	*US Citizens + Perm. Residents*	*% to Foreign Students*
1980	1538	1269	17.5
1985	1836	1432	22.0
1990	2100	1497	28.7
1995	2162	1624	24.9
2000	1989	1241	37.6
2003	1950	1262	35.3

Source: NSF Survey of Earned Doctorates/Doctorate Records

Limiting the discussion to the chemistry graduate school enrollments at the top ten National Research Council's (NRC) ranked departments yields some interesting observations. There are eleven schools included in this discussion because three departments, University of Chicago, University of Wisconsin-Madison and University of California at Los Angeles (UCLA), tied for 9th place in the ranking. As can be seen, there were wide variations in the size and gender distribution of the entering classes (Table IV). The private universities had much smaller enrollments than the large state universities. An extreme example is a comparison of Berkeley (University of California, Berkeley) and Columbia University where three times as many students matriculated at the former institution. There was also a wide difference in the percentage of women in the entering classes; the fraction of women in the entering graduate school classes at Harvard University (19.9%) was significantly lower than at UCLA (35.8 %).

Table IV. Chemistry Graduate School Enrollment at the Top Ranked NRC Departments, 1980-2002

Institution	*Total*	*Men*	*Women*	*% Female*
U.C., Berkeley	1,807	1,287	520	28.8
CalTech	902	666	236	26.2
Harvard	747	598	149	19.9
Stanford	832	588	244	29.3
MIT	1,085	732	353	32.5
Cornell	838	563	275	32.8
Columbia	559	373	186	33.3
U. Ill. at Urbana-Champaign	1,329	936	393	29.6
U. Wisconsin, Madison	1,172	828	344	29.4
U. of Chicago	634	474	160	25.2
UCLA	1,113	715	398	35.8

Source: NSF-NIH Survey of Graduate Students & Post-doctorates in S&E.

A cursory examination of the total number of doctorates granted in chemistry by all schools shows that there was a rise in the degrees awarded that peaked in 1970 at 2,214 that was followed by a slow decline and then by another surge in 1995 (Table V). On close examination of the doctorate pool, it is clear that dramatic changes were occurring in both the citizenship and gender composition of the recipients. A breakdown of the doctorates by citizenship shows that in the 1990s the increase in the number of doctorates awarded to foreigners and domestic women contributed significantly to the second surge and also dampened the later fall off. The growth in size of those two groups was critical to domestic, academic research efforts, as universities would have been hard pressed to maintain their competitive edge in light of the dwindling numbers of male students from the US.

Focusing on changes in the U.S. citizens and permanent residents' pool, it can be seen that women have retarded the decline in the total number of domestic doctorate recipients. The decrease in the number of domestic male doctorates coupled with the rise in domestic female doctorates resulted in women earning 34% of the doctorates in 2002 *(2)*. There was a slight retrenchment in 2003 as 31.5% of the degrees were earned by women. The slow growth in the number of women obtaining doctoral degrees means that it will take many years for the Ph.D. pool to mirror the ~50% female composition of the bachelor pool.

Previously, we determined for each gender the percent of entering graduate students that subsequently received doctorates, a value that we called the "yield" *(4)*. Using data obtained from the NSF WebCASPAR database, we ascertained for both genders the number of the first-time, full-time graduate students from 1988 to 1992 at a given school and the number of the doctorates granted from 1994 to 1998 by that chemistry department. To calculate the yield, the sum of the doctorates awarded to a gender was divided by the sum of the entering graduate students for that gender. In order to compare the performance of women and men, the yield for the women was divided by that for the men to give a value called the "parity index". A parity index value of less than 1 would indicate that the doctorate attainment rate for the women had been lower than that for the men.

Determination of the yield and parity index for the top 25 NRC ranked departments (University of California, San Francisco was not included because of its specialized curriculum) gave an average female yield of 62% and an average parity index of 0.85. This means that 62% of the entering female graduate students received a doctorate during the five-year window and that this percent was 85% of the value for the male students. Broadening the number of schools analyzed to include all PhD-granting schools, the yield for the female students was calculated to be 47% and the parity index was 0.77. Clearly, at both groups of schools women were not doing as well as the men in obtaining doctorates.

At the top 10 NRC departments, the female yield was 69% with a parity index of 0.89, indicating that the female students fared much better at this elite

Table V. Composition of the Chemistry Doctorate Pool, 1966-2002

		All Genders		*Men*		*Women*	
Year	*Total*	*U.S. Citizens + Perm. Residents*	*Temp. Residents*	*U.S. Citizens + Perm. Residents*	*Temp. Residents*	*U.S. Citizens + Perm. Residents*	*Temp. Residents*
1966	1,556	1,379	177	1,307	158	72	19
1970	2,214	2,038	176	1,882	152	156	24
1975	1,735	1,521	214	1,367	180	154	34
1980	1,506	1,269	237	1,051	204	218	33
1985	1,762	1,432	330	1,154	265	278	65
1990	2,027	1,497	530	1,114	421	383	109
1995	2,116	1,624	492	1,087	377	537	115
2000	1,888	1,241	647	819	474	422	173
2003	1950	1262	688	847	488	415	200

Source: NSF Survey of Earned Doctorates/Doctorate Records.

group of schools than at the lower ranked institutions. However, it should be pointed out that there was a wide range in both the female yield (42.9-85.3 %) and parity (0.60-1.04) values (Table VI). The female yield at Cornell University (82.8%) and at Chicago (85.3%) was substantially higher than that at the University of Illinois, Urbana-Champaign (42.9%) and UCLA (47.9%). In comparison to the doctorate attainment rate for the male students, the female students at MIT (Massachusetts Institute of Technology), Columbia, and Chicago did as well as the males and at Illinois they did much poorer.

In graduate school in chemistry, students select a focus for their research. The principal sub-specialties of the doctorates in 2001 were organic (26.3%), analytical (16.8%), physical (14.5%) and inorganic chemistry (14.1%) *(5)*. the attractiveness of the various sub-specialties has varied through the years in response to employment trends; however, men and women have responded similarly to the job market (Table VII) *(6)*. In 2001, there was little difference in the percent of men and women graduates majoring in physical or inorganic chemistry. The greatest gender difference was observed in organic chemistry, which consistently has had a higher percentage of men than women.

Table VI. Comparison of Doctoral Attainment Rates at the Top Ranked Chemistry Departments

Institution	*Male Yield*	*Female Yield*	*Parity Index*
U.C., Berkeley	79.6 %	67.7 %	0.85
CalTech	68.9	54.8	0.80
Harvard	82.6	72.7	0.88
Stanford	70.8	65.7	0.93
MIT	64.5	64.7	1.00
Cornell	87.3	82.8	0.95
Columbia	75.3	75.6	1.00
U. Ill., Urbana-Champaign	71.0	42.9	0.60
U. Wisconsin, Madison	91.3	74.7	0.82
U. of Chicago	82.2	85.3	1.04
UCLA	54.5	47.9	0.88

Source: NSF Survey of Earned Doctorates/Doctorate Records.

Table VII. Percent of Doctorates in Chemistry Sub-Fields, 1960-2001

	Organic		*Analytical*		*Physical*		*Inorganic*	
Year	*Men*	*Women*	*Men*	*Women*	*Men*	*Women*	*Men*	*Women*
1960-4	41.9	36.4	6.9	9.4	28.6	27.9	9.2	11.8
1970-4	36.6	28.5	7.7	5.2	23.2	26.2	13.1	16.0
1980-4	31.1	25.8	13.5	11.4	17.8	20.3	12.2	14.3
1990-4	24.6	21.0	13.4	15.6	16.2	16.1	11.4	12.4
1995-9	26.4	22.5	15.2	18.1	15.1	13.1	12.4	12.9
2001	28.8	21.0	15.7	19.4	14.4	14.6	13.9	14.9

Source: Survey of Earned Doctorates collected by the National Organization for Research at the University of Chicago.

Table VIII. Work Place for PhD Chemists

	All chemists			*Men*			*Women*		
Work Sector	*1990*	*1995*	*2000*	*1990*	*1995*	*2000*	*1990*	*1995*	*2000*
Industry	49.1%	49.8%	52.4%	50.3%	51.9%	54.3%	40.2%	38.9%	43.6%
Government	8.0	7.9	7.1	7.9	7.6	7.0	8.9	9.6	7.6
Other Non-academic	7.1	4.6	4.7	7.1	4.4	4.7	7.4	5.6	4.9
Academic	35.8	37.6	35.8	35.8	36.1	34.0	43.5	45.8	43.9

Source: The 1990, 1995 and 2000 American Chemical Society censuses of working members

Employment

Over the past decade, there have been slight gender changes in the work sectors where doctoral chemists have been employed (Table VIII) *(6)*. Men with doctorates were more likely to work in industry than were women; whereas, women were more often than men to be employed in academe. In 2000, 53% of the men and 43.6% of the women worked in industry. Equal numbers of women worked in industry and academe, and in government and other non-academic positions, there were few gender differences.

Academe

Chemists in academe are employed at associate, bachelor, master, and doctoral-degree granting institutions, medical schools and high schools: with PhD-granting institutions being the major employer for both men and women with doctorates *(6)*. Through the years, the percentage of men working at doctoral institutions has been consistently higher than that for women (Table IX). It should be noted that in 2000, the percent of men (48.2%) and women (35.0%) was lower than in 1995 for both genders. In general, women have been increasing their fraction of faculties. In 2000, 34.4% of the faculty positions at associate degree-granting institutions were held by women, 29.0% at bachelor institutions, 25.2% at masters' institutions and 20.2% at doctorate institutions. In high schools, female chemists were 46.2% of the chemistry faculties.

Table IX. Type of Academic Institution Employing Chemists

		Men			*Women*	
Institution Type	*1990*	*1995*	*2000*	*1990*	*1995*	*2000*
AA-granting	6.3%	6.3%	6.8%	8.8%	9.9%	10.2%
BS-granting	21.0	19.2	21.1	23.6	20.6	24.8
MS-granting	9.8	9.5	9.7	8.1	8.9	9.4
PhD-granting	52.2	52.6	48.2	41.0	42.0	35.0
Medical School	7.2	8.5	9.4	7.9	8.2	8.6
High School	3.4	3.8	4.8	10.5	10.4	11.9

Source: The 1990, 1995 and 2000 American Chemical Society censuses of working members.

PhD-granting Institutions

Another way of monitoring the progress that women are making in academe is to examine their percentage at the various professorial ranks. The fraction of faculty positions held by women at all ranks at bachelor's, master's-, and doctoral-degree granting institutions has consistently increased since 1980 (Table X) *(3)*. The percentage of women at the assistant level rank at master's-granting institutions (36.1%) and at bachelor's-granting schools (42.3%) is noteworthy; those values exceed the distribution of women in the doctoral pool (~30%).

At full, associate, and assistant ranks at PhD-granting institutions, women have continued to be represented below their distribution in the doctoral pool. This under-representation of women in tenure-track positions was occurring while women held in 2000 a disproportionately large number of the instructor/adjunct professor positions (45.7%) and their fraction in the other non-tenure track positions was also quite high: with women holding 27.0% of the research appointments, 34.5% of the non-faculty jobs and 31.5% of the positions with no rank *(6)*. Forty seven percent of the high school teachers were women.

The low percentage of female full professors at academic institutions certainly reflects the low numbers of women receiving doctorates in past decades. In turn, the higher percentage of women on faculties at the assistant and associate levels probably has resulted from both the increased societal pressure to hire women and the rise in the numbers of women obtaining doctorates. However, it should be pointed out that at the full professor rank, which is the most populous rank, the percentage of women is around 10%. In 2000, 54.9% of faculty members were at the rank of full professor, 21.7% at associate professor, and 23.4% at assistant professor.

The slow rate of hiring female faculty members by PhD-granting institutions has been noted by other researchers. The MIT study showed that the percentage of women on the faculty, 7%, had barely changed during the years 1985-94 *(7)*. In a series of reports, the ACS ascertained the number of faculty members along with their rank and gender for the top 50 federally funded chemistry departments *(8-12)*. During the years 2000-4 there has been very little change in the hiring of tenure-track female faculty members: the fraction of women on the faculty grew from 10.5% to 12.4% (Table XI).

The slow growth in the number of female assistant professors hired by the top federally funded schools is particularly striking considering that 3,192 women and 7,235 men *(13)* received doctorates in chemistry during the years 1997-2001. Granted that the ACS did not track the exact same departments each year because of slight changes in the rankings of top fifty funded schools; however, this data clearly shows the difficulty women were and are continuing to have in garnering tenure-track positions at top PhD-granting institutions. It is

Table X. Female Percentage of Faculty Members by Institution and Rank, 1980-2000

	PhD-granting			*MS-granting*			*BS/BA-granting*		
Year	*Full*	*Assoc.*	*Asst.*	*Full*	*Assoc.*	*Asst.*	*Full*	*Assoc.*	*Asst.*
1980	2.4%	5.2%	13.1 %	4.1%	10.2%	22%	7.6%	12.6%	20.3
1985	2.9	9.2	11.4	5.6	10.0	23.5	7.2	19.2	24.2
1990	4.3	12.2	18.4	5.9	14.2	28.8	9.5	23.4	30.7
1995	5.3	14.5	22.4	10.1	21.8	29.6	9.9	25.7	36.2
2000	7.9	18.0	25.2	11.2	25.6	36.1	13.7	32.0	42.3

Source: The 1980, 1985, 1990, 1995 and 2000 American Chemical Society censuses of working members.

Table XI. Rank of Female Faculty Members at the Top Fifty Federally Funded Chemistry Departments

	Full Professor		*Associate Professor*		*Assistant Professor*			
Year	*Women*	*Rank %*	*Women*	*Rank %*	*Women*	*Rank %*	*Total Faculty*	*Female Faculty*
2000	69	6.2	47	20.5	56	18.5	1641	10.5%
2001	74	6.7	46	19.7	61	20.3	1640	11.0%
2002	75	6.9	47.5	19.8	66	21.4	1630	11.6%
2003	87.5	8.2	46	20.4	63	21.2	1592	12.3%
2004	89	8.5	47	19.0	59	19.7	1594.5	12.4%

Source: Chemical & Engineering News.

interesting to note that the number of female associate professors has remained constant over these five years. In fact, the greatest growth occurred in the full professor rank, which went from 69 to 89 female faculty members.

Monitoring the gender changes in the faculty composition of the top 25 NRC ranked chemistry departments clearly shows the very slow progress that women are making in attaining tenure and tenure-track positions. Between the years 2000-2004, the number of female full professors increased from 37 to 48 at those schools, for female associate professors it improved from 17 to 23, and for female assistant professors there was a slight change, going from 26 to 28. The hiring of so few female assistant professors is rather shocking, as 821 women received doctorate degrees during the years 1994 to 2003 from a group of ten departments that in the past have trained a high percentage of faculty members at the top 50 NRC ranked universities. It should be noted that there was shrinkage in the overall number of chemistry faculty members at the 25 NRC schools, as the total number of faculty decreased from 595.5 to 558.

Possible Reasons for the Under-representation of Women

The slow rate of hiring of women by the top PhD-granting departments could be explained by a poor match in the sub-specialty chosen by women and the faculty positions that universities were trying to fill: simply, women had gone into sub-disciplines that were not in demand. Previously, we discussed the slight differences in the areas of specialization selected by men and women (Table 6). However, a comparison of the percent of women trained in organic chemistry and the fraction of female faculty members with that expertise shows that hiring does not reflect availability (Table XII) *(15)*. During the years 1960-98, 24% of women had selected organic chemistry as their sub-specialty; whereas, only 11% of the female faculty members in 1993 were organic chemists. The converse is true for physical chemistry, where 19% of the female doctorates had studied in that sub-field and 25% of the female faculty members were physical chemists.

Looking at the women at the assistant professor level, 21% were analytical chemists, 17% were inorganic chemists, 21% were physical chemists and 13% were organic chemists. In three of the sub-specialties, hiring by PhD-granting institutions had approximated the percentage of women trained in that expertise; in one sub-discipline, organic chemistry, insufficient numbers of women had been hired.

Another explanation for the low numbers of female faculty members at PhD-granting institutions could be that women had disproportionately attended schools that were not held in high esteem. Our research shows that this was not the case *(16)*. A comparison of the gender distribution of the doctorate pools from the top ten NRC schools and from all PhD-granting institutions showed no difference: the percent of women in both groups was the same.

Table XII. Comparison of the Training and Hiring of Chemists by Sub-discipline and Gender, 1993

	PhD's 1960-98			*Sub-Discipline on Faculties*		
	Overall	*Men*	*Women*	*Overall*	*Men*	*Women*
Analytical	12%	11%	13%	12%	12%	11%
Physical	20	21	19	28	28	25
Inorganic	13	12	14	16	17	15
Organic	31	33	24	25	26	11

Sources: Professional Women and Minorities by the Commission on Professionals in Science and Technology, and American Chemical Society Directory of Graduate Research – 1993.

In order to ascertain if women receiving doctorates from the highly regarded departments were being hired at the same rate as men graduating from those same schools, we determined the doctoral school of the tenured and tenure-track faculty members at the NRC-50 schools in 2001 *(16)*. This discussion will be limited to the faculty members who had graduated during the period 1979 to 2000. We ascertained that a select group of universities had trained sixty percent of the faculty members. Those departments were: Berkeley, Caltech, Harvard, MIT, Stanford, Wisconsin, Columbia, Cornell, Chicago and Yale. Obviously, the graduates from those ten schools were preferred by the hiring committees at the NRC-50 institutions over graduates from other schools.

Further examination of the NRC-50 faculties showed that fourteen departments had trained three or more female faculty members (Table XIII). With the exception of Michigan, Penn State (Pennsylvania State University), and U.C. at San Diego, the remaining institutions were in the select group of ten departments.

To determine if the female doctoral graduates from the fourteen schools had been more or less successful than their male counterparts in obtaining faculty positions, a value called the "impact factor" was determined for each gender. This calculation was done by dividing the number of NRC-50 faculty members trained at a university (according to the 2001 DGR) who had received their doctorate after 1978 by the total number of doctorates granted by that department during the years 1979-2000 and then multiplying by 100. A high impact factor for a gender would be indicative of a large percentage of the male or female doctoral graduates from a particular school holding a tenured or tenure-track position on a NRC-50 faculty in 2001. In a similar calculation, an overall impact factor was determined using the total number of faculty members trained at a school and the total number of doctorates granted by that department. This latter factor permitted comparison of the universities on the basis of faculty placement of their doctoral graduates.

At five of the fourteen universities, the female impact factor was greater than that for the males and at one, Wisconsin; the impact factors were the same for both genders (Table XIII). The female doctorates at Berkeley, Cornell, UCLA, Michigan and Penn State had attained faculty positions at a higher rate than their male counterparts. At the remaining nine schools, the male graduates had fared better than the female graduates in attaining faculty positions: a most troubling finding, and one that bears closer study. Clearly, graduating from an institution held in high regard by hiring committees is not working as well for women as for men. In comparing the overall performance of their doctoral graduate, those from Caltech and Harvard had done substantially better in attaining NRC-50 faculty positions than graduates from the other schools, as indicated by their high impact factor values.

Table XIII. The Doctoral Departments of NRC-50 Female Faculty Members

NRC-Ranking	*Institution*	*Number Women Hired*	*Impact Factor*		
			Female	*Male*	*Overall*
1	UC, Berkeley	21	7.9	7.7	7.7
2	Caltech	11	9.1	11.8	11.2
6	Cornell	7	5.3	4.3	4.5
10.3	Wisconsin	6	3.1	3.1	3.1
3.5	Harvard	6	7.6	10.9	10.4
10.3	UCLA	5	3.2	1.7	2.2
7	Columbia	5	4.7	6.9	6.4
5	MIT	5	2.9	5.8	5.1
3.5	Stanford	5	4.1	9.9	8.6
35	Michigan	4	2.5	0.8	1.2
18.5	Penn State	3	2.0	0.6	0.9
18.5	UC, San Diego	3	2.7	3.1	3.0
10.3	Chicago	3	3.4	5.1	4.8
12	Yale	3	2.7	6.1	5.2

Sources: American Chemical Society Directory of Graduate Research-2001, NSF Survey of Earned Doctorates/Doctorate Records.

Another explanation for the low hiring rate of women by PhD-granting institutions is that women had decided disproportionately not to seek post-doctoral appointments, a prerequisite for positions at Research I institutions *(14)*. Similar to the pattern discussed previously with the doctoral pool, since 1980 the number of male U.S. citizens and permanent residents has been decreasing; whereas the number of female post-docs has increased (Table XIV). From 1980 to 2002 the number of female U.S. citizens and permanent residents holding post-docs increased by 150%. However, during that time the number of foreign post-docs grew by 180% and for female foreign post-docs it went up by 470%. The latter group significantly increased the percent of women holding post-doctoral positions.

The percent of post-doctoral positions held by US citizens and permanent residents has been in decline in recent years; indicating that if this trend continues, foreign students will dominate the future faculty candidate pools. The National Science Foundation's *Science and Engineering Indicators-2004* reported that on average 64% of the temporary residents who received their doctorate from an American university in 1996 where still residing in the U.S. during the years 1997-2001.

It is not an encouraging sign that the percent of post-doctoral positions held by women was only 22.2%: probably predicative that the growth in the number of female faculty members will be slow, especially for U.S. citizens and permanent residents. It should be remembered that the number of women holding a post-doctoral includes all women at the various years in their post-doc training. As has been observed before in this chapter, the same pattern of women dropping out rather than going to the next stage of training occurred at this level; the percentage of postdocs held by women is smaller than their fraction of the doctoral pool.

Summary and Discussion

At each stage of education, we have shown that a significant number of women have elected to conclude their studies in chemistry prior to completing their post-doctoral studies: thus precluding consideration for a faculty position at a Research I institution. A simple picture of the choices women have made during their education can be seen by examining the gender distribution of the pools that contributed to the 2002 post-doctoral pool. Using data from NSF, in 1994 women obtained 41% of the baccalaureate degrees in chemistry; six years later in 2000 they were 32% of doctoral pool, and in 2002 they held 22% of the

Table XIV. Composition of the Post-Doctoral Pool, 1980-2002

	Men		*Women*			
Year	*U.S. Citizen + Perm. Resident*	*Foreign Post-docs*	*U.S. Citizen + Perm. Resident*	*Foreign Post-docs*	*% of All Post-Docs Held by Women*	*% of All Post-docs Held by U.S. Citizens + Perm. Resident*
1980	986	1,400	217	117	12.3	44.2
1985	1,035	1,488	250	237	16.2	42.7
1990	1,056	1,988	271	332	16.5	36.4
1995	1,185	1,765	318	379	19.1	41.2
2000	1,051	2,053	304	469	19.9	34.9
2002	883	2,189	319	559	22.2	30.4

Source: NSF-NIH Survey of Graduate Students & Post-doctorates in S&E.

postdoctoral positions. Clearly, at every stage women have opted out of continuing their formal education at a faster rate than the men.

We found that, in comparing the performance of either gender in obtaining a doctorate degree, a lower percentage of female graduate students received a PhD than did the men. The "yield" for female doctorates was 77% of that for the men. However, at the top ten ranked departments, women fared much better in comparison to the men and the yield for the women was 89% of the yield for men. Examination of the performance of men and women at the top 25 ranked departments showed a wide school-to-school variation. The yield value for the women varied from 42.9 to 85.3%, clearly showing the importance of wisely choosing a graduate school. At several schools, women did as well as the men in obtaining doctorates, whereas at others, women did significantly poorer. This wide spread in the attainment of doctorates by women in comparison to men most likely reflects differences in the environment that women were experiencing. Other chapters in this book have discussed the treatment women reported that they received in graduate school. It is not a great leap of faith to believe that the feelings of isolation and low support experienced by the women resulted in their early departure from graduate school without the receipt of a PhD.

For the women who persisted in school and obtained a doctorate from one of the major suppliers of faculty members, the data shows that they did not fare as well as the men from the same department in obtaining tenure track positions at PhD.-granting institutions. The low rate of hiring of women can not be attributed to their specializing in less popular or "wrong" sub-disciplines, as the choices made by women and men were quite similar except for organic chemistry, which has consistently attracted fewer women. It is interesting to note that the women trained in organic chemistry have had greater difficulty in attaining faculty positions than women trained in the other major sub-specialties.

Women have been more successful in attaining faculty positions at institutions offering lesser advanced degrees. In 2000, women held 34.4% of the faculty positions at associate degree institutions, 29.0% at bachelor degree-granting and 20.2% at PhD-granting departments. At the top 50 federally funded departments, the percentage of female faculty members was even lower, 10.5%, than at doctoral schools taken as a whole. From 2000 through 2004, the percentage of female faculty members has slowly increased to 12.4% at the top 50 funded universities. A total of 3 additional female assistant professors were hired by the 50 schools during the 5-year span. It should be noted that this low rate of hiring of women by those research institutions occurred while a large number of women obtained doctorates from departments that have trained a high percentage of 50 NRC faculty members. During the years 1994-2003, 821 women received doctorates from that elite group of schools. It appears that very few of them were able to attain a tenure track position at a top 50 federally funded university. This low rate of hiring also happened while the fraction of women in the U.S. citizen and permanent resident doctorate pool was increasing. In contrast to the difficulty women have experienced in attaining tenure track positions at PhD-granting institutions, they are over-represented in non-tenure track positions.

During the past twenty years, there has been a rise in both the number of foreign graduate students and post-docs. This was particularly true for female foreign post-docs. While foreign representation was increasing, the number of male U.S. citizens and permanent residents in graduate school and holding post-docs has been declining. This latter trend will make it increasing difficult for hiring committees to identify suitably trained, male U.S. citizens in the future. It will be interesting to see if this trend will result in the hiring of more US women or international students for faculty positions.

A number of fields, which previously had very few female professionals, now have a significant fraction of their terminal degrees received by women. In 2002, women earned 72% of the 2,289 doctorate degrees in veterinary medicine, 66% of the 7,076 doctorates in pharmacy, 48% of the 38,981 J.D./ LL.B degrees in law, 44% of the 15,237 M.D.'s awarded in medicine, and 39% of the 4,239 D.D.S/D.M.D degrees in dentistry *(18)*. In comparison, chemistry granted 1922 doctorates in 2002 with 34% granted to women.

There has been a constant rise since the mid 1970s in the number of women receiving doctorates in medicine. For example in 1970, 699 women (8.4% of the total) received an M.D. degree; in 1985 that number had grown to 4874, and in 2002, 6,768 women received an M.D. degree. In comparison, the increase in the number of women receiving doctorates in chemistry has been much smaller. According to the NSF's WebCASPAR database, 182 (8.11%) women received doctorates in chemistry in 1970, 362 women in 1985, and in 2002, the number of female doctoral recipients had grown to 647. Chemistry had attracted one-tenth as many women as medicine and had not kept up with medicine's rapid rate of growth. Since 1979 the number of women receiving doctorates in medicine had gone up 970%; whereas in chemistry it had increased by 360%.

Why have women gone into the other scientific fields in such great numbers? Did they envision greater acceptance of women in those fields? What subtle message was sent to the female graduate students and the numerous female undergraduate chemistry majors, especially those at PhD-granting universities, where there was and continues to be a paucity of women holding tenured and tenure track positions? Did chemistry fail to project that it was welcoming of both men and women and that a woman could aspire to becoming a tenured professor at a research university? The answers to those questions are complex and not clearly understood. We do know from the survey of graduate students at the top ranked departments that women did not perceive a supportive environment. This could have significantly impacted a number of decisions those women made, such as continuing on in graduate school or switching to a different field or career.

Changes will have to occur in the education of graduate students so as to attract substantially larger numbers of women. We have shown that women in chemistry have not been hired by the top PhD-granting institutions in tenure track positions in proportion to the available pool. Women have gone to the "right" schools and not specialized in the "wrong" fields and yet they have not been hired by the Research I institutions. Instead growing numbers of women have decided to pursue advanced degrees in other professional fields. For the U.S. to maintain its technical edge, it is imperative that chemistry draw the best women and men to address the scientific problems of tomorrow. Over the past twenty years, fewer men born in the U.S. and male permanent residents have been receiving doctorates in chemistry, when will research institutions decide that women should be well represented on their faculties?

References

1. Honing, L.S. In *Equal Rites, Unequal Outcomes*; Editor L. S. Horning, Kluwer Academic/Plenum, NY, 2003, p.5.
2. Heylin, M., Chem. & Eng. News, February 7, 2005, 38-46
3. Kasper-Wolfe, J., "What the Survey of Earned Doctorate Tells Us About Women in Chemistry: Selected Variables by Sex", presented to the ACS Working Group on Data Collection on Women and Minorities Employed in the Chemical Related Sciences, August 18, 2003.
4. Kuck, Valerie J. *Chem. & Eng. News*, November 19. 2001, 71-73.
5. Kasper-Wolfe, J. American Chemical Society, unpublished results, Washington, D.C., 2003.
6. *Women Chemists 2000, Analysis of the American Chemical Society's Comprehensive 2000 Survey of the Salaries and Employment Status of its Domestic Members*, American Chemical Society, **2001**, 17.
7. Massachusetts Institute of Technology "A Study on the Status of Women Faculty in Science at MIT, Massachusetts Institute of Technology, Cambridge, MA, 1999.
8. Long, J.R.,. *Chem. & Eng. News*, September 25, 2000, 57.
9. Byrum, A., *Chem. & Eng. News*, October 1, 2001, 99.
10. Long, J.R., *Chem. & Eng. News*, September 23, 2002, 111.
11. Marasco, C.A., *Chem. & Eng. News,* October 27, 2003, 59.
12. Marasco, C.A., *Chem. & Eng. News,* , September, 2004, 27-33
13. National Science Foundation's Survey of Earned Doctorates/Doctorate Records. *Enhancing the Postdoctoral Experience for Scientists and*

Engineers, Committee on Science, Engineering, and Public Policy, National Academy Press, Washington, D.C., **2001**, 11.

14. Roscher, N.M. and Cavanaugh, M.A., *Communicator,* Council of Graduate Schools, Washington, DC, December 2000., 4-5.
15. Kuck, V.J., Marzabadi, C.H., Nolan, S.A., and Buckner, J.P.; J. of Chem. Ed., 2004, 81(3), 356-363.
16. *Enhancing the Postdoctoral Experience for Scientists and Engineers,* Committee on Science, Engineering, and Public Policy, National Academy Press, Washington, D.C., **2001**.
17. National Center for Education Statistics, Digest of Education Statistics, 2003, http://nces.ed.gov/programs/digest/index.asp.

Chapter 10

Women Leaving Science Jobs: With Special Attention to Chemistry

Anne Preston

Department of Economics, Haverford College, Haverford, PA 19041

The 30-year period from 1970-2000 was the setting for remarkable changes in the educational and labor force achievements of U.S. women. With the passage of Title VII of the Civil Rights Act in 1964 and Title IX of the Educational amendments in 1974, employment and educational discrimination against women were outlawed. Labor force participation of women increased from 43 percent to 60 percent, women became the majority recipients of bachelors and masters degrees, and the once "etched in stone" female to male weekly earnings ratio increased from 0.63 to 0.77. In the male dominated professions of science and engineering public policy designed to increase the number of women in the science and engineering pipeline augmented federal legislation and the percentage of bachelors, masters, and PhD degrees in the sciences awarded to women increased by 12 percentage points (*1*). This marked success in degrees granted to women in the sciences masked a growing problem in the workplace. The number of women leaving science after receiving a science education and starting a science job was high and rising. With special attention when possible to the field of chemistry, the following chapter gives estimates of the levels of occupational exit of women scientists, explores factors behind exit, and recommends workplace policies to address this problem.

Data

The career paths studied come from four complementary data sets. The first data set, the Survey of Natural and Social Scientists and Engineers (SNSSE), 1982-1989, was collected by the National Science Foundation and gives background data on exit for a national sample of working scientists (*2*). The survey, which asks questions concerning job, demographic, educational, and personal characteristics, was sent in 1982 to a stratified systematic sample of more than 100,000 respondents to the 1980 Census. The full sample included a potential sample of working scientists, individuals who in 1980 worked in a set of targeted "science related" occupations and had four or more years of college education, and a potential sample of working engineers, individuals who worked in occupations targeted as "engineering" and who had two or more years of college education. All respondents were resurveyed in 1984, 1986, and 1989. Of those surveyed in 1982, only respondents who were employed, who answered "yes" to the question "are you working in a position related to the natural or social sciences?" and whose stated occupation was in the natural sciences or engineering were tracked over time. The assumption is that these individuals were working in science in 1982. Occupational exit had occurred by 1989 if the individual was not employed or if he or she responded "no" to the questions about whether his or her position was related to social or natural sciences in 1989.

In the 1990s NSF's national data collection efforts refocused on SESTAT, a compilation of surveys, including The Survey of Doctorate Recipients, The National Survey of Recent College Graduates, and The National Survey of College Graduates, which aims to identify the science workforce as those with degrees in science and engineering, rather than those working in science (*3*). The SESTAT data have information on individuals in 1993, 1995, 1997 and 1999 and, like the earlier SSE, give the researcher the potential to track individuals over time. In the SESTAT data an individual was "in science" if he or she was working and if he or she identified an occupation which, according to date compilers, was included in computer and math sciences, life and related sciences physical and related sciences, or engineering. These major occupational groupings were recoded from 500 potential responses to a detailed occupational location question. As in the case of the SNSSE data, those respondents identified as "in" science in 1993, who also responded to the survey in the final year of the decade, became the sample for which exit is estimated. For this constrained sample, exit occurs if the individual is not working in a science job in the final year of the survey and can be due to unemployment, labor market departure, or work in non-science employment. Employees who retired or who left due to a disability are eliminated from the samples for both the SNSSE data of the 1980s and the SESTAT data of the 1990s. Although the two data sets differ in terms of sampling procedures and definitions of "in" and "out" of science, the identical methods used to estimate exit give roughly accurate national patterns of occupational departure from the sciences.

The third data set complements the first two by giving more in-depth information on careers of a set of relatively homogeneous individuals. These data are the result of a work history survey sent to the population of active female alumnae and a random sample of active male alumni who received degrees in science, math, or engineering from a large public university in the northeast from the time of its establishment in the mid 1960's until 1991 (*4*).[1] The survey, implemented between 1992 and 1994, asks questions with the goal of describing the complete educational, personal, and work force histories of the respondents since college graduation. "In" science is defined as in the SNSSE data, and follow-up questions ask about reasons for occupational exit. Occupational exit rates are only calculated for graduates with science degrees who actually begin work in science.

The fourth data set contains interview information from roughly 100 of the respondents to the work history survey and was designed to understand more fully the factors behind occupational exit of men and women in the sciences (*4*). Twenty six pairs of women from the original university sample were selected to participate in interviews concerning both their education and career experiences. From the willing respondents, the 52 women were initially selected to mirror the age, education, and family distribution of the respondents to the work histories.[2] Within each pair, the two women are similar in age, degree level, field of degree, and family circumstances. The difference between the two women in each pair is that one of the women has left science and one has stayed. The purpose of this pairing process is to help isolate the important factors behind exiting or continuing scientific careers that cannot be identified using standard statistical techniques. Twenty-six pairs of men were also identified and interviewed. The male pairs are matched to the female pairs so that individuals in the two pairs

---

[1] 5200 surveys were sent out, roughly 400 were returned due to out of date addresses, and 1688 were completed, for a response rate of 35%.

[2] Only 51 women were interviewed. The 52nd woman had died between the time in which she filled out a survey and the time of the scheduled interview. This woman had a PhD in Physics, and because of the small number of women with PhD's in physics, no similar women could be found.

have the same age, family characteristics, level of degree, and subject of degree.[3] Information from the interviews fill in the details, allowing a deeper understanding of the causes and consequences of occupational exit.

Magnitude of Exit from Science and from Chemistry

Table I gives estimates of exit using the three survey data sets. The 1980s national data sets show that in the seven years of the 1980's about 8.5 percent of the male science workforce left science while twice this percentage of women left. Compared to occupational exit rates for other occupations calculated using the 1987 Current Population Survey these exit rates are quite high. In particular over the one year period preceding the 1987 survey 0.6 percent of health diagnosing professionals and 0.7 percent of lawyers and judges left their occupations (*5*). Both annual estimates point to a 7 year exit rate below 5 percent. The relatively high exit rates of women from science during this period are a result of high rates of labor force departure (row 2, column 1) and high rates of exit to other occupations (row 2, column 3). Interestingly while the exit rates for male chemists exit are comparable to the rates for all male scientists, women in chemistry have lower exit rates than all women in science (rows 2 and 4, column 4). The big difference is exit to non- science occupations. A similar percentage of chemists and all female scientists are out of the labor force (rows 2 and 4, column 4), but less than 6 percent of female chemists, compared to over 9 percent of all female scientists, work outside of science.

These patterns in exit are only partly replicated in the university data. As in the SNSSE data, the exit rates for all female scientists (28.2% -- row 6, column 4) are approximately twice the exit rates for all male scientists (14.3% -- row 5, column 4). Exit rates are naturally higher for the university sample since, on average, respondents had been in the labor force for 12 years. Men in chemistry had lower exit rates than all male scientists but female chemists' exit rates were not significantly different than exit rates for all female scientists (rows 6 and 8, column 4).

By the 1990s occupational exit rates had risen markedly possibly due to technical personnel looking for the riches that the dot com boom seemed to

---

[3] Because of the differing field distributions of men and women where men are relatively over-represented in engineering and women are relatively over represented in biological sciences, there are three pairs of men which have different subject areas than their female counterparts.

promise. The largest increases in exit were for men who according to interview data are more likely than women to leave for increased salary and opportunity. Over the period from 1993 to 1999, 19 (row 9, column 4) percent of the male scientists had left science and 17.5 percent left to work outside of science (row 9, column 3). Female scientists' exit rates increased to 29.1 percent for the six year period (row 10, column 4), and in comparison to the 1980's the big increase was in the percent of women scientists working outside of science. Female chemists' exit rates (16.2% -- row 12, column 4) continued to be lower than all female scientists' (29.2% --row 10, column 4) exit rates in the national estimates, and surprisingly female chemists were exiting science at roughly the same rate as male chemists.

Table II gives occupational exit rates for PhD scientists estimated with the same samples. As expected occupational exit rates should be lower for scientists with PhD's since they have invested in high levels of education. The educational investment itself is a signal of commitment, and the resulting high level of skills and knowledge and the salary they can command create high costs of leaving science. Most interesting is the fact that the gap between male and female exit rates disappears once one focuses on the PhD scientists. While for most groups of female PhDs there is still a small percent who leave the labor force (column 1), presumably to care for family, the percentage working outside of science is small. These patterns extend to female chemistry PhDs where none of the 1980s sample (row 4, column 3), none of the university sample (row 8, column 3) and only 7.6% of the 1990's national sample (row 12, column 3) were working outside of science.

Tables I and II reveal that occupational exit rates for scientists are high and rising. Women are more likely to leave than men except at the PhD level where in some instances women are more tied to science than men. Generally women chemists are more attached to science than women in other science fields, and women with PhD's in chemistry have very high attachment rates. In the university sample, similar to men, all the women who earned a PhD in chemistry were working in science. PhD drop out rates, however, show a slightly different picture. Table III reveals that PhD drop out rates are more than twice as high for women (32.3%) than for men (14.5%). Drop out rates from chemistry PhD programs are higher than in all sciences for both men and women in the university sample with 36.8 percent of women and 18.2 percent of women leaving chemistry PhD programs. Although once the PhD has been achieved occupational exit is similar for men and women, the road to the PhD, whether in chemistry or other sciences, creates more serious roadblocks for women than for men.

Table I. Exit Rates of All Natural Scientists and Chemists

	Exit from the Labor Force (1)	*Exit to Unemployment* (2)	*Exit to Non-science Employment* (3)	*Total Exit* (4)
	1980s National Sample (7 year period)			
1. Male Scientists (n=17070)	0.5	2.1	6.1	8.7*
2. Female Scientists (n=2468)	6.5	1.7	9.2	17.4**
3. Male Chemists (n=560)	0.7	1.6	6.1	8.4*
4. Female Chemists (n=239)	6.7	0.8	5.9	13.4
	1970-1994 University Sample (12 year period)			
5. Male Scientists (n=519)	1.7	1.7	10.8	14.3*
6. Female Scientists (n=623)	12.0	2.2	14.0	28.2
7. Male Chemists (n=29)	3.4	0.00	6.9	10.3*
8. Female Chemists (n=44)	11.4	0.00	13.6	25.0
	1990s National Sample (6 year period)			
9. Male Scientists (n=17624)	0.7	0.9	17.6	19.2*
10. Female Scientists (n=4191)	6.0	1.4	21.7	29.1**

Table I. *Continued*

	Exit from the Labor Force	*Exit to Unemployment*	*Exit to Non-science Employment*	*Total Exit*
	(1)	*(2)*	*(3)*	*(4)*
	1990s National Sample (6 year period)			
11. Male Chemists (n=1562)	0.6	1.3	14.1	16.0
12. Female Chemists (n=369)	6.8	1.6	10.8	16.2

* Total exit rate for males is significantly different than total exit rate for females at the .01 level.

** Total exit rate for all females is significantly different than total exit rate for female chemists at the .01 level.

Table II. Exit Rates of PhD Natural Scientists and PhD Chemists

	Exit from the Labor Force (1)	*Exit to Unemployment* (2)	*Exit to Non-science Employment* (3)	*Total Exit* (4)
1980s National Sample				
1. Male PhD Scientists N=2077	0.4	0.7	3.0	4.1
2. Female PhD Scientists N=366	1.6	0.5	1.6	3.7
3. Male PhD Chemists N=217	0.9	0.5	5.5	6.9
4. Female PhD Chemists N=43	4.7	0	0	4.7
1970-1994 University Sample				
5. Male PhD Scientists[a] (n=116)	0.9	0.00	7.8	8.7
6. Female PhD Scientists (n=108)	1.9	0.9	4.6	7.4**
7. Male PhD Chemists (n=18)	0.0	0.0	0.0	0.0
8. Female PhD Chemists (n=12)	0.0	0.0	0.0	0.0
1990s National Sample				
9. Male PhD Scientists (n=8254)	0.6	0.8	12.8	14.2*

Table II. *Continued*

	Exit from the Labor Force (1)	*Exit to Unemployment* (2)	*Exit to Non-science Employment* (3)	*Total Exit* (4)
		1990s National Sample		
10. Female PhD Scientists (n=1859)	3.6	1.2	14.0	18.8
11. Male PhD Chemists (n=1246)	0.5	1.5	12.8	14.8
12. Female PhD Chemists (n=236)	5.9	1.3	7.6	14.8

[a] The exit rate from science is calculated for those PhDs who started careers in science. 2.5 % of the male PhDs and 4.4 percent of the female PhDs never entered a science job.

* Total exit rate for males is significantly different than total exit rate for females at the .01 level.

** Total exit rate for all females is significantly different than total exit rate for female chemists at the .01 level.

Why Do Scientists Leave

This question is not an easy one to answer and must be addressed with the differing complementary data sets.

Statistical Analyses

Statistical analyses of the probability of leaving science using the national and university data are helpful in identifying correlations between individual or work characteristics and the probability of leaving science. The results of the analyses, which are consistent across all data sets, identify educational, field, and personal characteristics which affect the probability of leaving in a marginal manner-- tipping the scales a little bit one way or another towards staying or leaving.

Table III. Drop-out Rates from PhD Programs—University Sample

Drop Out Rates from PhD Programs:	*Males*	*Females*
All Science (n=305)	14.5%*	32.3%
Chemistry (n=41)	18.2%*	36.8%

*The male percentage is significantly different from the female percentage at the 0.01 level.

Most noticeable is the effect that level of education and type of education have on the probability of exit. Exit rates vary by major of degree and by level of degree most likely because of differences in the extent to which different educational degrees train recipients for a job or a career. In this context the distinction between engineering degrees and science degrees is important. Relative to a bachelor degree in science, which develops a body of knowledge and a way of thinking that can be helpful in a variety of careers, the engineering degree, regardless of level, is a more narrow, professional degree, one that prepares the graduate for a job and a career specifically in engineering. Of the graduates who never spent a day working in science, 90 percent were science majors while only 10 percent studied engineering. Likewise, the exit rate of engineering majors was roughly half the exit rate of science majors. As noted earlier respondents with PhD degrees were less likely to leave science possibly because the skills learned in a PhD program, regardless of whether they are in a science or engineering field, are directly transferable to the jobs and career paths expected of this elite group of scientists and because those individuals selecting to pursue a PhD are the same men and women who are the scientists most talented and excited about the fields.

The characteristics associated with the scientific field from which the worker graduated influence exit. Multivariate analysis of the national data show that the probability of exit decreases in fields where salaries are increasing. In addition in fields where the rates of growth of knowledge are accelerating, scientists are more likely to leave, possibly to avoid the increasing amounts of retraining and skill update that such a fast changing field requires. Characteristics of the worker's situation also affect exit. A scientist whose salary is below average, for scientists with his or her characteristics, is more likely to leave than one with an average or above-average salary. Part time workers are more likely to leave science than those on a full-time schedule. Being married significantly reduces the probability of exit for all reasons for men and has no significant effects on the probability of exit for women except in the instance of exit from the labor force where marriage increases that probability significantly. Having children increases the probability of exit to other occupations for both men and women. However, women with children are more likely to leave the labor force than their childless counterparts, while the opposite is true for men.

Responses to Survey Questions

In the retrospective work histories, those individuals employed outside of science at the time of the survey were asked specifically why they had left science. Each respondent had the opportunity to cite at most three reasons for exit from the sciences. The results, presented in Table IV, show that men overwhelmingly focused on the low pay in science jobs (68%) and the lack of opportunities to advance (64%). However, in decreasing order of importance, they also cited other fields being more interesting (36%), the lack of science and engineering positions (34%), a preference for non-science positions (23%), and promotion out of science and engineering (18%). While low pay and lack of science opportunities were also important to women with roughly a third of the women citing each of these reasons, a large number of women also identified: a preference for other positions (35%), other fields are more interesting (30%), a lack of science and engineering positions (21%), the difficulty of having a family and working in science and engineering (21%), length of hours required of a science and engineering position (20%), and unfriendliness of the science and engineering fields to women (19%). Thus while men exit science, overwhelmingly because of a lack of opportunities and low pay, women leave for multiple reasons.

Interview Data

Interview data focus even more specifically on the overriding reason behind exit. As noted earlier, the interview sample of 104 men and women chosen from the 1688 respondents to the work history survey, was constructed as a set of pairs where individuals differed only according to whether they had exited or stayed in science. The results from the interviews are very similar to the results from the survey responses. Table V gives a summary of the results. The men left science primarily to find better career options in terms of higher pay and better advancement opportunities. Of the 19 pairs of men for whom a primary (and sometimes secondary) factor differentiating the pairs could be identified, 15 (79%) of the men left in response to salary or career opportunity. In contrast, only in one pair of women was the desire for greater pay and more promising career opportunities the major differentiating factor behind the leaver and the stayer.

For women the reasons behind their decisions to exit were more varied, and three important reasons for exit surfaced. In eight of the twenty-two pairs of women for whom a primary factor differentiating the stayer from the leaver could be identified, the reason for exit was a mismatch of interests. The woman

Table IV: Reasons Why Men and Women Left Science

(Work History Sample: n=1688)

Percent who cited:	*Men*	*Women*
Pay better in non-science and engineering positions	68.0	33.0
Career opportunities lacking	64.0	34.0
Other fields more interesting	36.0	30.0
Science and engineering position not available	34.0	21.4
Preferred other Positions	23.0	35.0
Promoted out of science	18.0	2.9
Impossible to have a family and work in S&E	4.5	21.4
Demands of the career are too severe	4.5	2.9
Hours required too long	0	20.0
Science and engineering unfriendly to women	0	19.0

Source: Reproduced with permission from reference 4. Copyright 2004 Russell Sage Foundation.

who stayed found the scientific field interesting relative to other opportunities while the women who left did not. A mismatch of interests was also a primary or secondary factor differentiating the men in nine of the pairs. In seven of the twenty-two pairs of women, the positive guidance of a strong mentor was the primary difference between the women who stayed and those who left. Finally, family responsibilities were the major factor behind occupational exit in six (less than a third) of the twenty-two pairs of women.

The construction of the interview sample, which was designed to identify a factor differentiating the leaver from the stayer, necessarily downplays some important issues that affect careers of scientists regardless of whether they stay or leave. In particular, to conclude that men's preoccupation with money and career swamps family concerns would not be fair to these men, many of whom see income and career growth as the best way to provide for their families. In addition many men who stay in science do so because their concerns for the security and stability of their families prevent them from undergoing risky career moves. The relatively small number of women in the interview sample who leave science because of family concerns does not mean that family issues were easily solved by women who were balancing work and family. Rather almost every woman was grappling with these issues so that it was not a factor that could differentiate many stayers from leavers. Similarly, perceptions by women of sex discrimination and double standards were prevalent among the interviewed women. However sex discrimination and double standards were only secondary factors in exit decisions as they contributed to low levels of mentoring, a mismatch of interests, and difficulties in shouldering the double burdens of family and career. Furthermore many of these women had dealt with sex discrimination since high school and had found strategies to persist in science in spite of unequal treatment.

More on the Factors Causing Women to Leave Science

Discontent with Science

The one common thread in all the discussions with men and women who leave science because of discontent with the field was the narrowness of science. Many scientists found the work itself very narrow and specialized, while other exiting scientists, especially those at the PhD level, expressed concern that in order to succeed in science the scientists themselves have to become very narrow. Women, more than men, were dissatisfied with the isolation associated with working in science; unhappy that the work involved little personal contact

Table V. Factors Differentiating Leaver From Stayer in Interview Pairs

	Men	*Women*
Discontent with Income and Opportunity in science	15	1
Primary Factor	0	1
Secondary Factor		
Looking for More Interesting Work Outside of Science	3	8
Primary Factor	6	1
Secondary Factor		
Lack of Mentor or Guidance		
Primary Factor	0	7
Secondary Factor	0	1
Difficulty Shouldering Familial and Career Responsibilities		
Primary Factor	1	6
Secondary Factor	0	1
Number of Pairs for which a Factor Differentiating Leaver from Stayer could be Identified	19	22

with colleagues either within or outside of science and had no connections to real life issues that they found important.

Not only were women more likely than men to exit science because of discontent with science, of those individuals remaining in science, women were more likely than men to voice concerns about the nature of the work. This discrepancy might occur for a variety of reasons. Women may be more likely to act in response to this disillusionment with science because they are less likely to be the family bread winner and have more freedom than men to leave a job if it does not work out. Without an exit option, men may be less likely to criticize the work that they perform. Alternatively the typical structure of a man's scientific career may result in a broadening of responsibilities that satisfies expanding interests. Women's careers may remain more narrow because of lesser opportunities and more non-work obligations. Finally, women, because of social and familial roles, may have different expectations about paid work. Astin (*6*) notes that when students were asked about career interests, women were more likely than men to answer that their future work would contribute to society, help others, give them the opportunity to work with people and ideas and to express themselves. While it is possible that all these factors contribute to the high number of women leaving science in response to discontent, it is also true that in the interview data women leaving for this reason entered careers with nurturing components such as high school teacher, lay minister and clinical psychologist.

Family Formation and Responsibilities

Familial responsibilities affect career outcomes in very different ways for men and women as traditional roles of women continue to exist even in this population of highly educated scientists and engineers. Many men talked about their early careers as a time when they put in lots of work hours to earn higher salaries and greater career opportunities; they were specializing in work for the sake of family at the same time that women were compromising work for the sake of family. The two to one ratio of female to male exit rates extended to many of the familial patterns. According to the work history data women were roughly twice as likely as men to marry a spouse with an advanced degree, have a full time working spouse, and sacrifice own career for spouse's career. Responses to question concerning responsibility for household chores and childcare for preschool children which are displayed in Table VI reveal that men and women agreed that the woman of the family took on roughly two thirds of household chores and childcare responsibilities while the man took on only a sixth of the responsibility for childcare. For women, the most common response to the difficulties of shouldering family and career responsibilities was not occupational exit but a lower level career compromise such as limiting job search to a specific geographical location, working part-time, or forgoing promotions that required travel. In the interview data, eighty percent of married women with children and fifty percent of childless married women engaged in this type of career compromise.

Table VI. Family Responsibilities by Sex

	Women	*Men*
% of household chores spouse is responsible for	34.8* (649)	65.1 (479)
% of child care spouse is responsible for	15.1* (449)	67.0 (363)
% of child care individual is responsible for	60.2* (449)	17.6 (363)

* Percentage for women is significantly different than percentage for men at the 0.01 level.

Numbers in parentheses give the sample size for which the percentage was calculated.

Source: Reproduced with permission from reference 4. Copyright 2004 Russell Sage Foundation

Interestingly the most difficult balancing acts occurred at different stages of family formation for men and women with different levels of education and career aspirations. For PhD scientists the career track, which requires early geographical mobility as the developing scientist moves from undergraduate institution to graduate institution to first post doctoral position to first job, puts stress on the dual career couple as they navigate this career route along with the often conflicting career route of the professional spouse. Not only is the academic scientist moving frequently early in the career, but academic institutions in which he or she locates are often located in rural settings where land is inexpensive and work opportunities outside the institution are sparse. Because female scientists are more likely to be part of a dual career couple than male scientists, they are more likely to feel this stress. Not all relationships succeed but in those that do, career compromise is a necessity, and because women are likely to be the younger, less established partner in a relationship, they are likely to be the ones who make the compromise. Every married woman in the interview sample narrowed the geographic scope of her job search to accommodate her husband's career. Many of these women felt that they did not have the right to ask their husbands and families to move for their jobs, and they preferred not to carry the added responsibility and potential guilt associated with such a move.

Women who terminated their science education with a bachelors or masters degree found that the greatest difficulty balancing work and family comes with the birth of children. These women had more plentiful opportunities than their counterparts with PhDs since their education is less specialized and they can turn to private industry or government as well as the nonprofit sector for employment. Therefore job location for the dual career couple was not usually a stumbling block to the relationship or the career. Meeting the conflicting demands of young children and an often inflexible work environment, however, posed difficult hurdles to these women. Some women expected that one parent would stay at home with the pre-school child, but most looked for a work environment which allowed a comfortable shouldering of the double burden of work and family. Opportunities for part-time work, flexible hours, minimal travel and overtime hours, and no relocation requirements fit the bill. Generally it was the woman who sacrificed career opportunity for comfort either because the husband was more established and his participation would generate large financial sacrifices or because the woman felt that the family sphere was her responsibility.

Lack of Mentoring

From comments of respondents to the interview sample, the role of mentor in both men's and women's careers is less as a role model or one who inspires and more as one who teaches and shows the way. In a productive mentoring relationship there is a definite transfer of human capital from mentor to mentee. Therefore it should not be surprising that mentoring stands out as an extremely important factor influencing career decisions and dictating career outcomes of science educated women in the university sample. Mentoring early in the science career had an immediate impact on the woman's probability of continuation and success in science. On the other hand, mentoring of men, while more prevalent, especially in the academic arena, had a less pronounced effect on short-term career outcomes. The apparent differences in extent and impact of mentoring for men and women is not surprising since science is a male dominated field. Mentoring relationships may develop naturally for men because a large majority of the potential mentors in science are men, but the guidance may be knowledge that the men could gather from interactions with peers. At the same time female scientists may perceive the science workplace as a foreign landscape where any guidance is helpful. Table VII reveals the stark differences in mentoring experienced by the male and female interview respondents. As undergraduate science students, 40 percent of the men reported having a mentor compared to only 13.5 percent of the women. The difference was amplified in graduate school where two thirds of the men and only one fifth of the women reported having a mentor. As noted earlier the effect of mentoring was much greater for women. Only 60 percent of the women without mentors graduated while every women mentored as a graduate student earned the graduate degree. The probability of graduating for men was not affected by mentoring. Men and women were equally likely to be mentored in an early employment situation, largely because mentoring in industry is the result of institutionalized programs. But again the effects of mentoring were much greater for women. An early employment outcome was defined to be positive if the individual experienced salary or opportunity growth, while the outcome was defined as negative if the job was dead-end, the individual's career stagnated, the individual was laid off, or the job led to scientific exit. All the women who had been mentored had positive employment outcomes early in the career while only 52 % of the women without mentors had positive employment outcomes. Mentoring increased the probability of a positive outcome for men (from 70 % to 83 %) but by smaller amounts.

Table VII. Mentoring Statistics by Sex (Interview Sample--- n=102)

	Women	*Men*
% with mentor as an undergraduate	13.5%	40.0%
% with a mentor as a graduate student	20.5%	65.7%
Difference in probability of earning graduate degree "No mentor" to "With mentor"	0.6 to 1.0	No change
% with a mentor in nearly employment	52%	51%
Difference in probability of successful early employment outcome "No mentor" to "With mentor"	0.52 to 1.0	0.70 to 0.83

Graduate school mentors were often but not always the graduate advisor. The description of the graduate advisor varied more considerably for women than men where advisors of female graduate students ranged from inspiring teacher and facilitator to hostile "anti-mentor". While there were only three women in chemistry PhD programs in the interview sample, each one described their advisor as antagonistic and destructive. Two of the women left their PhD programs with a masters degree because of the advisor. Both women went into industry. The one who found a mentor in her industry job has a successful career as a chemist in a biotech firm. The other eventually quit to teach high school. The third woman changed fields within chemistry and eventually graduated with a PhD. Of the four men who started a PhD program in chemistry, all four described having mentors and three of the four earned the PhD. The difference in mentoring extends to the undergraduate years. Of the four women who where undergraduate chemistry majors, none had mentors as undergraduates, while 3 of the 5 men who were undergraduates in chemistry had mentors. These sample sizes are too small to make any general conclusions but they do call for further exploration of mentoring of women in chemistry programs.

Policy Recommendations

Women are more likely to leave science than men and they leave for different reasons. The reasons are largely due to alienation from work, from colleagues, and from potential mentors. In some sense this is due to the fact that science is a field developed by men for men. There are policy prescriptions that can be implemented both within and outside the academy that can go a long way to helping women succeed in science.

To deal with the dual career issues that often cause female PhD scientists to compromise their careers, science departments and university administrations should work together to find employment opportunities for spouses of desired job candidates. More importantly the requirement that PhD scientists make a number of geographical moves in the early stages of the career as they learn from different scientists in graduate school and post doctoral appointments must be reexamined, especially in light of the technological advancements in travel and communication over the last forty years. Offering policies designed to improve the quality of life of working parents, such as maternity/paternity leave, increased flexibility of work hours, telecommuting, unpaid personal days for childhood emergencies, a temporary part-time work option, and on-site daycare may not be enough to ease the conflicting burdens of children and work. Upper level management in these employing institutions must support these policies with credible promises that there will be no reprisals in return for taking advantage of childcare benefits.

Mentoring programs for all scientists should be set up and institutionalized in both academic and non-academic science workplaces. Because of the paucity of women in some fields, mentors to women need not be women themselves. However establish female networking programs in universities may be desired in addition to more general mentoring programs so that female scientists can create networks of female colleagues who are from a variety of scientific disciplines and at varying levels of the career. These colleagues can offer support, guidance, and friendship throughout the career.

Good career counseling for degree recipients in the different scientific disciplines is likely to reduce exit due to discontent with science since expectations about the job will be better informed. Better mentoring and networking of female scientists will also ameliorate feelings of isolation. Finally the trend toward inter-disciplinary work, which has taken place in the last twenty years, gives the individual scientist the opportunity to choose areas of work where the science itself can be connected to a bigger picture. University administrations must find ways to value and reward these initiatives which run counter to the more traditional disciplinary based research. All of these initiatives are likely to reduce occupational exit rates of both women and men. In addition by making the science workplace more hospitable and welcoming, they will have the extra advantage of attracting more individuals to these fields.

References

1. National Center for Educational Statistics, *Digest of Educational Statistics,* **2001**.
2. *Survey of Natural and Social Scientists and Engineers, 1982-1989*, Bureau of the Census, U.S. Department of Commerce.
3. *SESTAT Public Use File, 1993-1999*, Division of Science Resource Statistics, National Science Foundation.
4. Preston, Anne, *Leaving Science*, Russell Sage Foundation, New York, 2004.
5. Markey, James P. and William Parks II., "Occupational Change: Pursuing a Different Kind of Work," *Monthly Labor Review*, September 1989, *112*(9), pp3-12.
6. Astin, Helen S. "Patterns of Women's Occupations," in J. Sherman and F. Denmark, eds. *The Psychology of Women: Future Directions*, New York: Psychological Dimensions, 1979.

Indexes

Author Index

Subject Index

E

F

R

S

T

U

V

W